MECHANICAL TESTING
OF ENGINEERING
MATERIALS

...

Professor Kyriakos Komvopoulos
Department of Mechanical Engineering
University of California—Berkeley

cognella
San Diego, CA

First published in the United States of America in 2011 by Cognella, a division of University Readers, Inc.

Trademark Notice: Product or corporate names may be trademarks or registered trademarks, and are used only for identification and explanation without intent to infringe.

15 14 13 12 11   1 2 3 4 5

Printed in the United States of America

ISBN: 978-1-60927-920-2

www.cognella.com 800.200.3908

CONTENTS

Preface

The mechanical response of materials to different external loadings is of great importance to many fields of science, engineering, and industry. Structural failure is realized when the functionality of engineering components has been depleted. In general, there are three main reasons for a component to become dysfunctional—excessive (elastic or inelastic) deformation, fracture, and wear. Excessive elastic deformation is controlled by the elastic properties of the material, such as elastic modulus, and may occur under loading conditions of stable equilibrium (e.g., excessive deflection of a beam), unstable equilibrium (e.g., buckling of a column), and brittle fracture. Excessive inelastic deformation depends on plastic material properties, such as ultimate tensile strength, strain hardening, and hardness, and may occur under loading conditions conducive to fatigue (a process involving alternating stresses (or strains) that induce crack initiation from stress raisers or defects in the material followed by crack growth), ductile fracture due to excessive accumulation of plastic deformation, and creep (a time-dependent deformation process encountered with viscoelastic materials and elastic-plastic materials at elevated temperatures subjected to a constant stress). Material degradation may also occur as a result of mechanical wear arising at contact interfaces of load-bearing components when the transmitted contact stresses are comparable of the material hardness. It is therefore important to not only know how the mechanical properties control the material response to a certain external force, but also have knowledge of standard mechanical testing methods for measuring different material properties.

Chapter 1 is devoted to the modification of the material microstructure by heat treatment. The chapter shows how equilibrium phase diagrams and time-temperature-transformation diagrams can be used to predict steel microstructures after a certain type of heat treatment. Chapter 2 is focused on hardness measurement and the interpretation of the hardness data in terms of microstructure differences. Chapter 3 provides insight into the extraction of elastic and plastic material properties of different materials from uniaxial loading experiments. Chapter 4 treats fracture toughness measurement by impact testing and correlation of toughness differences to microstructure variations induced by heat treatment.

The objective of Chapter 5 is the mechanical behavior of viscoelastic materials, particularly microstructure effects on stress-strain and stress relaxation responses. Deformation due to cyclic loading is the theme of Chapter 6. Special attention is given to analyzing the data in the context of cyclic hardening (softening), plastic shakedown (constant plastic strain accumulation per cycle), and ratcheting (continuous accumulation of plasticity). The purpose of Chapter 7 is to illustrate how to perform strain-control and stress-control fatigue tests and how to correlate the measured fatigue properties to microstructure characteristics. Wear testing is the theme of Chapter 8, particularly the recognition of different lubrication regimes, seizure mechanisms, lubricant behavior, and effect of load, temperature, and sliding speed on friction and wear properties of bearing materials.

The sequence of the experiments included in this book is typical of most books covering mechanical behavior of materials. A set of questions is included at the end of each chapter, which can be used to design laboratory assignments. A report template and a rubric for evaluating laboratory reports are included in Appendices A and B, respectively. It is presumed that the students who use this book have had elementary courses on mechanics and materials sciences and, therefore, are familiar with such concepts as force and moment equilibrium, stress, strain, steel microstructures, and phase diagrams. They should also be familiar with basic statistics and data acquisition and processing.

This book should be a valuable supplement to an undergraduate-level course on mechanical behavior of materials. The emphasis is on mechanical testing according to established standards and the development of comprehensive reports of the obtained measurements. In addition to the necessary background and representative results, the book provides essential steps for carrying out each test. The coverage includes experiments drawn from the most important topics of an undergraduate course on mechanical behavior of materials.

—K. Komvopoulos
Berkeley, California

Acknowledgments

The idea for this book emerged after teaching the undergraduate course on mechanical behavior of materials at the University of California at Berkeley for more than 20 years, and recognizing the lack of understanding of basic concepts in the absence of hands-on experience and difficulties of the students to use classroom material to explain experimental findings in a comprehensive manner. The book reflects the author's strong commitment to promote excellence in undergraduate teaching at UC Berkeley and to elevate student understanding of materials behavior.

This book would not have been possible without the tireless efforts of many of my graduate research students, who took precious time from their research to assist me in developing the experiments presented in this book. I am greatly indebted to H. Zhang, H. Xiang, N. Wang, X. Yin, Q. Cheng, Z. Song, H. Xu, and A. Poulizac who not only played pivotal roles in setting up the experiments and obtaining the representative results included in this book, but also supervising the students during the laboratory sessions over the past year and grading their reports.

I also wish to acknowledge the support and encouragement provided by my Department. I especially thank Professor A. P. Pisano, Chair of the Department of Mechanical Engineering, for his continued interest on this project and for raising the funds to renovate the space where these laboratories are hosted. I am also grateful to my colleagues Professors L. A. Pruitt and H. Dharan for donating an axial loading machine and an impact tester, respectively, for the experiments and their support throughout my efforts to create these laboratories. Lastly, my thanks go to Mr. S. McCormick for technical assistance, equipment service and upgrading, specimen fabrication, laboratory coordination, and, especially, for overseeing the work of my students during the development of the experiments.

Chapter 1: Heat Treatment

1.1. BACKGROUND

Some basic information about phase diagrams and time-temperature-transformation (TTT) diagrams of steels is presented first as background material for the heat treatment part of this lab. Information for heat treatment of various carbon steels can be found in Appendix 1A.

1.1.1. Phase Diagrams

The equilibrium phase diagram of Fe-C alloy system is shown in Fig. 1.1 [1].

At eutectoid position (0.76 wt% carbon composition, 748°C), the alloy undergoes the following phase transformation:

γ (austenite) ↔ α (ferrite) +Fe₃C (cementite)

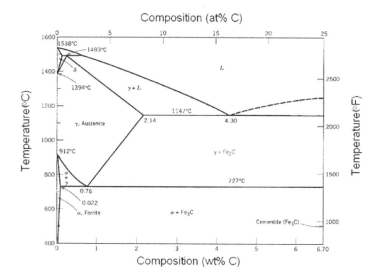

Fig. *1.1. Equilibrium phase diagram of carbon steel [1].*

If the composition is exactly the eutectoid composition (Fig. 1.2(a)), solid-solid phase transformation will occur upon cooling below 748°C, resulting in the formation of pearlite (Fig. 1.2(b)) [1]. However, if the composition is hypoeutectoid (Fig. 1.3(a)), then α phase will form first, while cementite will form after cooling below 748°C, resulting in a mixture of pearlite and α ferrite (Fig. 1.3(b)) [1].

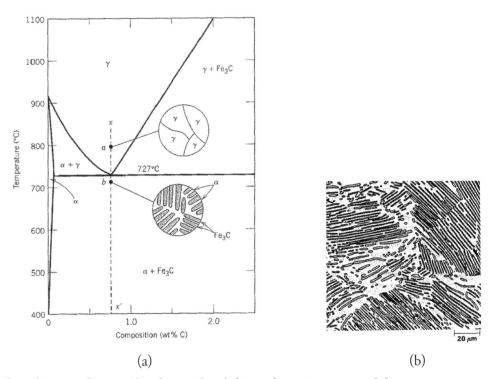

(a)　　　　　　　　　　　　　　　　　　(b)

Fig. *1.2. (a) Phase diagram of eutectoid carbon steel and (b) pearlite microstructure [1].*

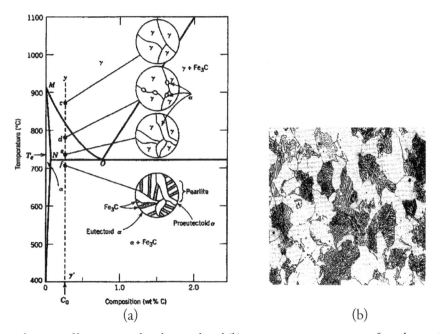

(a)　　　　　　　　　　　　　　　(b)

Fig. *1.3. (a) Phase diagram of hypoeutectoid carbon steel and (b) microstructure consisting of pearlite and* α *ferrite [1].*

In the case of hypereutectoid composition (Fig. 1.4(a)), cementite will form first, whereas α ferrite will form upon cooling below 748°C, producing a mixture of pearlite and cementite (Fig. 1.4(b)) [1].

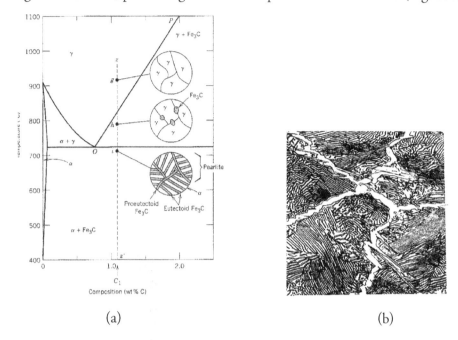

(a)   (b)

Fig. *1.4. (a) Phase diagram of hypoeutectoid carbon steel and (b) microstructure consisting of pearlite and cementite [1].*

1.1.2. Time-Temperature-Transformation (TTT) Diagrams

Phase transformation during heat treatment is also controlled by the cooling rate. At a fixed temperature, it takes a certain time for phase transformation to be completed. By plotting the phase transformation begin time and the 50% and 100% complete times as functions of time, the three curves shown in Fig. 1.5 [1] can be obtained.

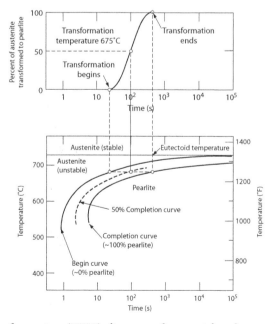

Fig. *1.5. Time-Temperature-Transformation (TTT) diagram of eutectoid carbon steel [1].*

Figure 1.6 shows the time-temperature-transformation (TTT) phase diagram of eutectoid carbon steel. If cooling from above 720°C is rapid (e.g., less than 5 s), the material does not have enough time to go through equilibrium phase transformation. Instead, the austenite phase transforms to a metastable martensite phase. This is called martensitic phase transformation [1].

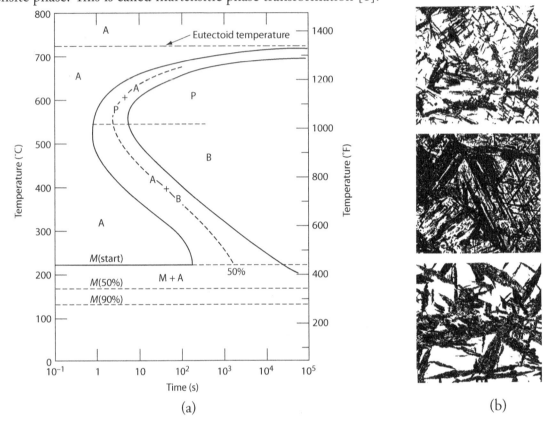

(a)

(b)

Fig. *1.6. (a) Time-Temperature-Transformation phase diagram of eutectoid steel and (b) steel microstructures formed at different cooling rates.*

1.2. EXPERIMENTAL

Sample preparation requires polishing (Fig. 1.7(a)), observation with an optical microscope (Fig. 1.7(b)), heating in an environmental-control furnace (Fig. 1.7(c)), and quenching the heat-treated samples in oil or water.

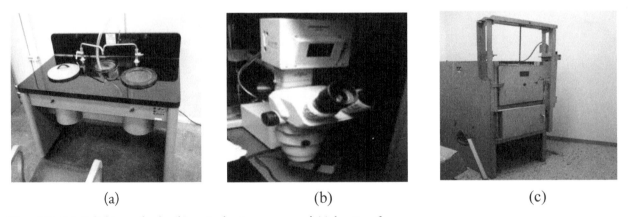

(a)   (b)   (c)

Fig. *1.7. (a) Polishing wheels, (b) optical microscope, and (c) heating furnace.*

1. Steel specimens consisting of AISI 1018 and AISI 1045 carbon steel will be polished sequentially with 320- and 600-grit SiC polishing paper. The better the steel sample is polished, the easier to observe the microstructure with the optical microscope.

2. Apply a chemical etchant (see details in Sect. 1.4.3), typically, 5 parts nitric acid and 100 parts alcohol, onto the polished steel surface to reveal the original microstructure.

3. Load the sample onto the microscope stage and observe the microstructure of the original (untreated) steel. Make a sketch of the expected microstructure and label the most characteristic features.

4. Heat the steel specimens to 850°C in the environmental furnace, hold the specimens at this temperature for 1 h, and then quench in oil (or water). Record the medium temperature and estimate the time for the specimen to reach the temperature of the quenching medium. Observe and describe the quenching process.

5. Repeat the previous steps for the other steel specimens.

Figure 1.8 shows AISI 1018 and AISI 1045 steel specimens before and after the above mentioned heating and oil-quenching processes.

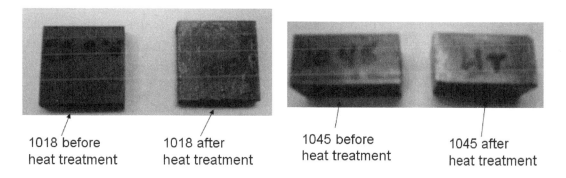

1018 before heat treatment   1018 after heat treatment   1045 before heat treatment   1045 after heat treatment

Fig. *1.8. Surface morphologies of steel specimens before and after heating at 850°C for 1 h in an environmental furnace and then quenching in oil.*

1.4. OPTICAL MICROSCOPE OBSERVATION OF METALLURGICAL MICROSTRUCTURE

The major steps are: sample mounting, polishing, chemical etching, and imaging.

1.4.1. Sample Mounting

To facilitate polishing, specimens should be sectioned to small pieces because small areas are easier to be polished flat than large areas. Epoxy mounting is highly suggested because the small specimen can be mounted into a large epoxy piece that is easy to hold by hand. The detailed procedure for sample mounting is as following:

1. *Sample cutting.* This step makes specimens into small dimensions, probably 5 mm diameter range.

2. *Preparation grinding.* This step grinds away the roughness by the previous cutting. It does not require very high smoothness.

3. *Heat treatment.* This step will produce the martensite phase in the steel alloy.

4. *Pre-etching.* Martensite phase transformation requires quenching, and quenching may cause oxidization on the steel surface. This step etches away the oxide scale(s).

5. *Cleaning*. Water cleaning of the steel surface.
6. *Mold preparation*. Prepare the mold for epoxy and place the specimen inside the mold.
7. *Weigh epoxy*. Epoxy needs to be mixed by precise formula; this step prepares the epoxy materials.
8. *Mix epoxy*.
9. *Encase specimens*. Pour mixed epoxy into the mold.
10. *Furnace treatment of the epoxy*. This step cures the epoxy. The result will be one big piece that is easy to hold by hand.
11. *Mold release*. This step uses a liquid medium to remove the epoxy-bound specimen out of the mold.

1.4.2. Polishing

The detailed procedure for polishing is as following:
1. *Belt grinder*. This is an electrical grinding facility that grinds away large asperities and surface contaminations by epoxy.
2. *Bench grinder*. This facility rolls the sand papers by electrical motor, with adjustable flow of lubricant. Usually four sand papers (240-, 320-, 400-, and 600-grit) are placed parallel on the bench grinder in order to save time from changing the sand paper. Epoxy-mounted specimens are progressively polished with finer sand paper.
3. *Additional bench grinder*. Because diamond paste polishing may not be effective after polishing with 600-grit sand paper, additional sand papers might have to be used to polish the sample. Typically three sand papers (800-, 1000-, and 1500-grit) can be used sequentially.
4. *Diamond paste polishing*. After sand paper polishing, a diamond paste (typically 6 or 1 μm particle size) is used to further polish the specimen surface. Usually Plexiglas plates (typically 8 x 8 in²) holding the diamond paste are used as the flat surface for polishing. Since 1 μm is close to the wavelength of visible light, using 1-μm-size diamond particles is effective for obtaining a final mirror-polished sample surface.

1.4.3. Chemical Etching

Nital and Picral are two typical etchants for alloy surface etching. Etching must be done immediately after polishing. Because nital (mixture of 2 wt.% of nitric acid in ethanol) can potentially explode, it must be stored with the cap loosely open, or with an inserted tube to allow for pressure release. The entire etching process must be done under a fume hood to maintain good air conditioning of the room and to prevent spill off. The typical etching steps are:
1. *Swab etching*. In this step, cotton floss sticks dipped with natal are used to swab the specimen surface.
2. *Water cleaning of specimen surface*.
3. *Alcohol cleaning of specimen surface*. This step is used to remove the water.
4. *Air drying*.

1.4.4. Imaging

Imaging with an optical microscope must be done immediately after chemical etching. The optical microscope has an objective lens of 40X and an eye piece of 10X; therefore the magnification is around 400X. Since the magnification scale is not automatically set by the microscopes, a 1-mm long calibration scale can be placed under the microscope along with the specimen to label the dimension of the specimens. A digital camera is used to obtain digital images when visual examination of the microstructure is satisfactory. It may be necessary to go back and forth to polishing, etching, and imaging until a satisfactory image is obtained.

1.5. REFERENCES

[1] Callister, W. D., *Materials Science and Engineering: An Introduction*, John Wiley, New York, 2003.

[2] Chandler, H., *Heat Treater's Guide*, 2nd ed., ASM International, Metals Park, OH, 1995.

[3] Van der Voort, G. F., *Atlas of Time-Temperature Diagrams*, ASM International, Metals Park, OH, 1991.

1.6. ASSIGNMENT

The main purpose of this lab assignment is to teach you how the microstructure of materials (in this case low-carbon (AISI 1018) and medium-carbon (AISI 1045) steel can be altered by heat treatment followed by rapid cooling (quenching). This requires knowledge of phase transformations induced due to the temperature variation.

Your analysis and discussion should address and provide explanations to the following:

1. Determine whether the steel specimen can be classified as hypoeutectoid or hypereutectoid.
2. Discuss possible phases present in the original (untreated) and treated steel specimens based on equilibrium phase diagrams and TTT diagrams provided herein.
3. Determine possible phases present in each steel microstructure before and after heat treating and quenching and justify your assignment.

1008 Isothermal Transformation Diagram

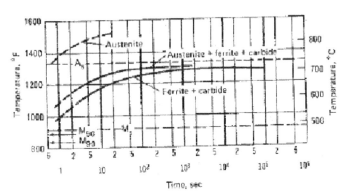

Heat Treater's Guide, 2nd ed

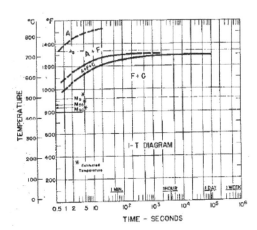

Atlas of Time-Temperature
Diagrams

Reference [2]

Reference [3]

1019 Isothermal Transformation Diagram

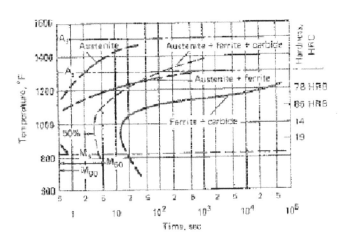

Heat Treater's Guide, 2nd ed

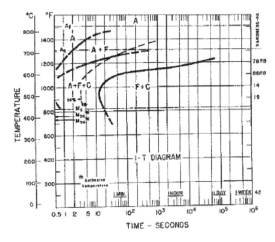

Atlas of Time-Temperature
Diagrams

Reference [2]

Reference [3]

1010 Time-Temperature Diagram

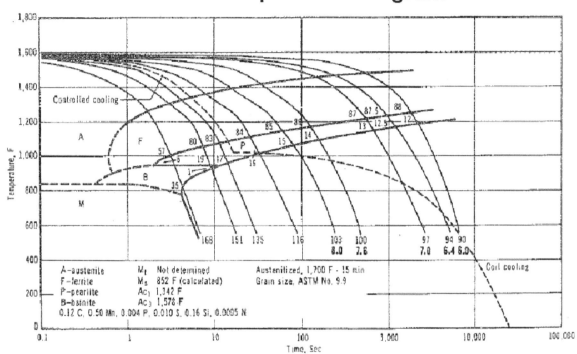

Heat Treater's Guide, 2nd ed

Reference [2]

1021 Isothermal Transformation Diagram

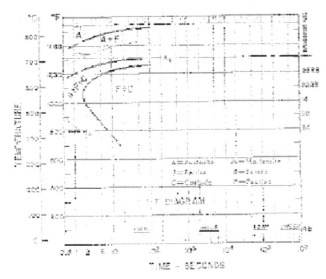

Heat Treater's Guide, 2nd ed

Reference [2]

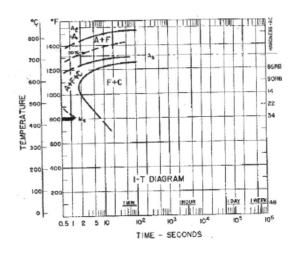

Atlas of Time-Temperature Diagrams

Reference [3]

1038 Cooling Curve

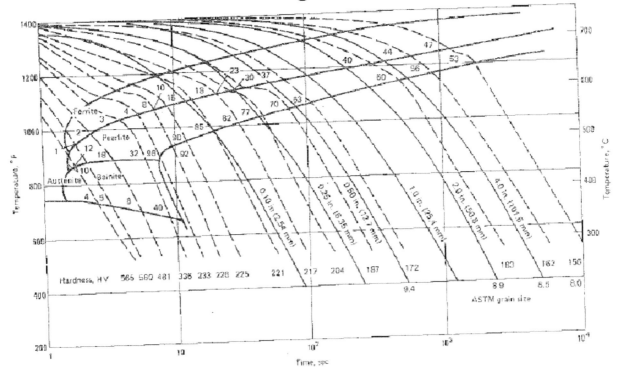

Reference [2]

1040 CCT Diagram

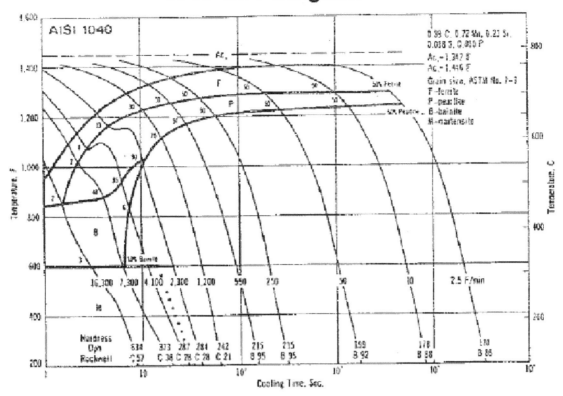

Heat Treater's Guide, 2nd ed

Reference [2]

1045 TTT Diagram

Ac1: temperature at which austenite begins to form
Ac3: temperature at which transformation of ferrite to austenite is complete

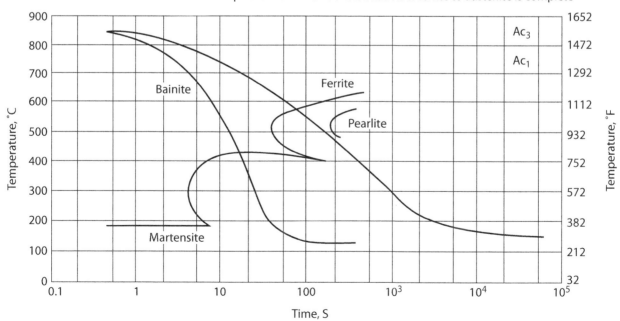

Reference [2]

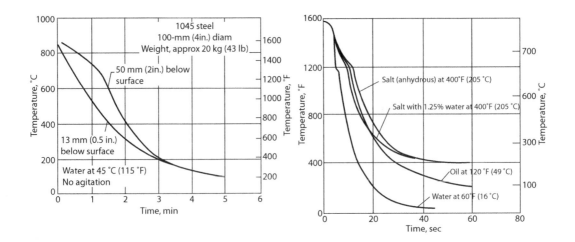

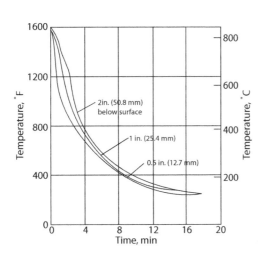

Reference [2]

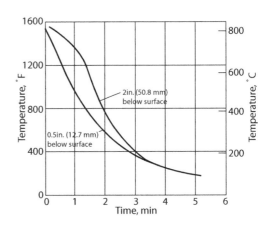

Reference [2]

Chapter 2: Indentation Hardness

cold work
grain refinement
solid soln strengthening
precip hardening
strain aging

2.1. BACKGROUND

Hardness is defined as the ability of a material to resist plastic deformation due to indentation by a rigid body. Generally, the material hardness is measured by pressing a sharp indenter of known geometry and mechanical properties against the test material.

The hardness of the material is quantified using one of a variety of scales that directly or indirectly indicate the contact pressure involved in deforming the test surface. Since the indenter is pressed into the material during testing, hardness is also viewed as the ability of a material to resist compressive loads. The indenter may be spherical (Brinell test), pyramidal (Vickers and Knoop tests), or conical (Rockwell test). In the Brinell, Vickers, and Knoop tests, hardness is the load supported by a unit area of the indentation, expressed in kilograms of force per square millimeter (kg_f/mm^2). In the Rockwell tests, the depth of indentation at a prescribed load is determined and is then converted to a hardness number (without measurement units), which is inversely related to the depth. Although all indentation hardness tests generally serve the same purpose, each test has certain advantages that make it more applicable to certain types of materials and part geometries. Brinell is used primarily on forgings and cast iron. Vickers and Knoop tests are used on very small or thin parts and for case depth determinations on parts such as gear-tooth profiles. The Rockwell test is the most popular indentation hardness test and is used in a wide variety of applications.

The Rockwell hardness test is defined in the ASTM E 18 standard as well as several other standards (Table 2.1) [1–4]. The Rockwell hardness number is based on the difference of the indenter depth from two load applications (Fig. 2.1). The minor load is applied first and a set position is established on the dial gauge or displacement sensor of the Rockwell tester. Then, the major load is applied. Without moving the piece being tested, the major load is removed and, with the minor load still applied, the Rockwell hardness number is determined by the dial gauge or digital display. The entire procedure takes from 5 to 10s.

In the Rockwell hardness test, the indenter may be either a diamond cone or a hardened ball depending on the characteristics of the material being tested. Diamond indenters are used mainly for testing relatively hard materials, such as hardened steels and cemented carbides. Steel ball indenters of diameter equal to 1/16, 1/8, 1/4, and 1/2 in. are generally used for testing materials such as soft steel, copper alloys, aluminum alloys, and bearings steels. Designations and typical applications of various indenters are given in Table 2.2.

Typically, the minor load in Rockwell hardness testing is set equal to 10 kg_f. No Rockwell hardness value is expressed by a number alone. Instead, a letter is assigned to each combination of load and indenter,

Table 2.1. Selected Rockwell hardness test standards.

Standard No.	Title
ASTM E 18	Standard Test Method for Rockwell Hardness and Rockwell Superficial Hardness of Metallic Materials
ASTM E 1842	Standard Test Method for Macro-Rockwell Hardness Testing of Metallic Materials
ASTM E 140	Standard Hardness Conversion Tables for Metals Relationship Among Brinell Hardness, Vickers Hardness, Rockwell Hardness, Superficial Hardness, Knoop Hardness, and Scleroscope Hardness
ASTM A 370–07	Standard Test Methods and Definitions for Mechanical Testing of Steel Products

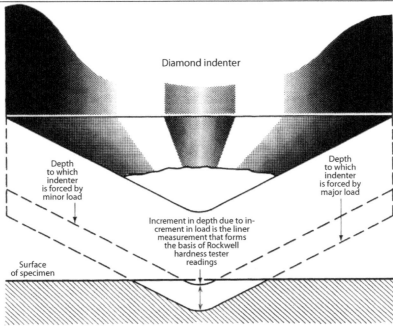

Fig. *2.1. The principle of Rockwell hardness testing. Although a diamond indenter is illustrated, the same principle applies for steel ball indenters [5].*

as indicated in Table 2.2. Each number is suffixed first by the letter H (for hardness), then the letter R (for Rockwell), and finally the letter that indicates the scale used. For example, a value of 60 in the Rockwell C scale is expressed as 60 HRC. One Rockwell number represents an indentation of 0.002 mm (0.00008 in.). Therefore, a reading of 60 HRC indicates an indentation from minor to major load of (100–60) × 0.002 mm = 0.080 mm or 0.0032 in. A reading of 80 HRB indicates an indentation of (130–80) × 0.002 mm = 0.100 mm [6].

2.2. SPECIMEN DIMENSIONS

The metal surrounding the impression in a Rockwell hardness test is considered to be "cold-worked." The depth of material affected during testing is on the order of ten times the indentation depth. Therefore, unless the thickness of the metal being tested is at least ten times the indentation depth, the accuracy of Rockwell hardness test will be questionable.

Table 2.2. Rockwell standard hardness [5].

Scale symbol	Indenter	Major load, kg$_f$	Typical applications
A	Diamond (two scales—carbide and steel)	60	Cemented carbides, thin steel, and shallow case-hardened steel.
B	1/16 in. (1.588 mm) ball	100	Copper alloys, soft steels, aluminum alloys, malleable iron.
C	Diamond	150	Steel, hard cast irons, pearlitic malleable iron, titanium, deep case-hardened steel, and other materials harder than 100 HRB.
D	Diamond	100	Thin steel and medium case-hardened steel and pearlitic malleable iron.
E	1/8 in. (3.175 mm) ball	100	Cast iron, aluminum and magnesium alloys, bearing metals.
F	1/16 in. (1.588 mm) ball	60	Annealed copper alloys, thin soft sheet metals.
G	1/16 in. (1.588 mm) ball	150	Phosphor bronze, beryllium copper, malleable irons. Upper limit 92 HRG to avoid possible flattening of ball.
H	1/8 in. (3.175 mm) ball	60	Aluminum, zinc, lead.
K	1/8 in. (3.175 mm) ball	150	Bearing metals and other very soft or thin materials. Use smallest ball and heaviest load that do not produce anvil effect.
L	1/4 in. (6.350 mm) ball	60	Bearing metals and other very soft or thin materials. Use smallest ball and heaviest load that do not produce anvil effect.
M	1/4 in. (6.350 mm) ball	100	Bearing metals and other very soft or thin materials. Use smallest ball and heaviest load that do not produce anvil effect.
P	1/4 in. (6.350 mm) ball	150	Bearing metals and other very soft or thin materials. Use smallest ball and heaviest load that do not produce anvil effect.
R	1/2 in. (12.70 mm) ball	60	Bearing metals and other very soft or thin materials. Use smallest ball and heaviest load that do not produce anvil effect.
S	1/2 in. (12.70 mm) ball	100	Bearing metals and other very soft or thin materials. Use smallest ball and heaviest load that do not produce anvil effect.
V	1/2 in. (12.70 mm) ball	150	Bearing metals and other very soft or thin materials. Use smallest ball and heaviest load that do not produce anvil effect.

In addition to the indentation depth limitation for a given specimen thickness and material hardness, there is a limiting factor of minimum specimen width. If the indentation is placed too close to the edge of the specimen, the Rockwell hardness will be underestimated. For accuracy, the distance from the center of the indentation to the edge of the specimen must be at least 2.5 times the indentation diameter. Multiple readings cannot be taken at the same point of a specimen surface. To ensure an accurate test, the distance from center to center of indentations should be apart by a minimum of three indentation diameters [6]. Minimum thickness requirements and conversions between various Rockwell scales are given in tabulated form in ASTM E 18 and ASTM E 140 [1–3] and have been reproduced in Appendix 2A.

Another basic requirement of the Rockwell hardness test is that the specimen surface must be approximately normal to the indenter apex and the specimen should be firmly fixed so that it does not move or slip during indentation. Because one regular Rockwell hardness number represents a vertical movement of the penetrator of approximately 0.000080 in., a movement of only 0.001 in. could cause an error of over 10 Rockwell numbers. The support must be of sufficient rigidity to prevent permanent deformation during indentation.

2.3. SCALE LIMITATIONS

Diamond indenters should not be calibrated for hardness values of less than 20, i.e., hardness readings below this level should be rejected. In the case of soft materials, results may not agree when the indenters are replaced, and another scale—for example, the Rockwell B scale—should be used from the onset.

Although scales that use a ball indenter (for example, the Rockwell B scale) range up to 130, readings above 100 are not recommended, except under special circumstances. This is because between ~100 and 130 only the tip of the ball is used. In view of the relatively blunt indenter apex, the sensitivity of most scales is poor in this region.

2.4. OPERATION PROCEDURE

Figure 2.2 shows a picture of the Rockwell hardness tester. Numbers correspond to important test components. The following steps should be followed in performing a hardness test.

1. Before starting a test, ensure that the crank handle (1) is pulled forward counter-clockwise as far as it can go. This lifts the power arm and the weights.
2. Select the proper indenter (2) and insert the plunger rod.
3. Place the proper anvil (3) on the elevating screw.
4. Place the specimen or test block on the anvil.
5. Bring the indenter in contact with the specimen by turning the capstan hand-wheel (4). Continue the motion until small pointer (5) shows that a *minor load* is applied.
6. Turn the bezel of the dial gauge to set dial zero behind pointer.
7. Release the weight (*major load*) by tripping the crank handle. *Do not force this crank handle. Allow the dashpot to control the speed of testing.*
8. When the large pointer (6) comes to rest, return the crank handle to the starting position. This removes the *major load*. However, the *minor load* is still applied.
9. Read the scale letter and Rockwell number from the dial gauge.
10. Remove the *minor load* by turning the capstan hand-wheel (4) counter-clockwise to lower the elevating screw and specimen such that they clear the indenter.
11. Remove the specimen and/or repeat the test.

2.5. REPRESENTATIVE RESULTS

Rockwell hardness tests were performed on seven calibration test blocks and six steel specimens. Hardness data from these tests are given in Table 2.3. Figs. 2.3 and 2.4 show the actual values of B and C scale calibration test blocks versus the experimental values. For the scale calibration test blocks, a linear relationship

Fig. *2.2. Rockwell hardness tester.*

exists between the actual values and the experimental values. Best-fit lines are also shown in Figs. 2.3 and 2.4. This indicates that HRC_a and HRC_e follow the relationship:

$$HRB_a = 1.015 \times HRB_e + 1.115$$
$$HRC_a = 0.955 \times HRC_e + 4.693 \tag{2.1}$$

Experimental hardness values of the six steel specimens, scaled according to Eqs. (2.1), are given in Table 2.4.

Both tensile strength and hardness are indicators of the material's resistance to plastic deformation. A hardness test is occasionally used to obtain a quick approximation of tensile strength. Appendix 2B contains a table for converting hardness measurements from one scale to another or to approximate tensile strength. These conversion values have been obtained from computer-generated curves and are presented to the nearest 0.1 point to permit accurate reproduction of those curves. All converted Rockwell hardness numbers should be rounded off to the nearest integer.

The Rockwell hardness of the six specimens was converted to ultimate tensile strength and Brinell and Vickers hardness in kg_f/mm^2 and MPa units, as shown in Table 2.4. Appropriate scales were selected considering the scale limitations. Empty entries are because converted hardness values for these cases are not appropriate

The yield strength in tension is about 1/3 of the hardness. Thus, the yield strength of the material σ_Y can be approximated as

$$\sigma_Y = \frac{H_v}{c} \approx \frac{H_v}{3} \tag{2.2}$$

where c is a constant determined by geometrical factors, and usually ranges between 2 and 4. The yield strength estimated from Eq. (2.2) is given in Table 2.4. The results indicate that the hardness of hot-rolled steel 1045 is less than that of the cold-formed steel 1045. The hot-rolling process can improve the ductility of the steel.

Table 2.3. Rockwell hardness test results.

Sample	Scale	Test										Average of ten tests	Standard Deviation
		1	2	3	4	5	6	7	8	9	10		
Calibration Test Block 46.5	B	44.0	44.0	44.5	44.8	44.6	44.8	44.8	45.5	45.0	45.5	44.8	0.5
Calibration Test Block 93.1	B	88.2	90.0	90.2	90.0	90.2	88.5	90.0	90.2	90.0	90.5	89.8	0.7
Calibration Test Block 93.8	B	91.5	92.0	92.0	92.0	92.0	92.1	91.8	91.8	92.0	92.2	91.9	0.2
Calibration Test Block 94.1	B	90.5	92.0	91.8	92.0	91.5	91.8	91.8	92.2	92.0	92.2	91.8	0.5
Calibration Test Block 26.4	C	21.0	23.2	22.8	23.0	23.2	23.5	23.0	23.2	23.0	23.0	22.9	0.7
Calibration Test Block 48.1	C	44.5	44.0	45.0	45.2	45.0	45.8	45.5	45.2	45.8	44.2	45.0	0.6
Calibration Test Block 62.4	C	60.5	60.8	60.0	60.8	60.5	61.0	60.8	60.5	60.5	61.0	60.6	0.3
Steel 1018 HR NHT	B	72.0	74.2	74.8	74.8	75.5	76.0	78.0	73.0	76.2	76.5	75.1	1.7
Steel 1018 HR HT	B	103.5	104.0	108.0	105.5	105.5	102.0	105.8	106.5	105.0	106.8	105.3	1.6
Steel 1018 HR HT	C	27.7	27.5	30.0	25.2	28.2	32.0	33.0	32.5	32.2	31.0	29.9	2.5
Steel 1045 HR NHT	B	89.2	88.0	88.0	89.0	89.0	89.2	89.8	90.0	89.8	90.0	89.2	0.7
Steel 1045 HR NHT	C	7.0	8.0	7.0	7.0	7.5	6.5	7.0	5.0	8.5	8.0	7.2	0.9
Steel 1045 HR HT	C	49.0	52.8	50.2	52.8	51.8	47.8	51.0	52.5	54.5	48.5	51.1	2.1
Steel 1045 CF NHT	B	99.2	98.8	98.8	98.8	99.0	100.0	98.8	99.2	98.0	99.2	99.0	0.5
Steel 1045 CF NHT	C	17.5	20.0	21.0	20.0	18.5	20.0	20.0	19.0	20.0	18.0	19.4	1.0
Steel 1045 CF HT	C	56.0	57.0	56.2	51.5	53.0	51.0	52.0	52.8	50.5	52.8	53.3	2.2

The hardness of the 1018 steel is less than that of hot-rolled or cold-formed 1045 steel. Steel 1018 has a carbon content of 0.15–0.20%, whereas steel 1045 has a carbon content of 0.43–0.50%. According to the test results, the higher carbon content resulted in higher hardness. This is mainly due to the presence of more alloying elements that introduce more dislocations, which increase the material resistance to

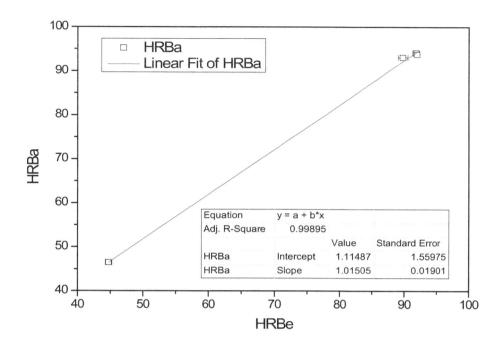

Fig. *2.3. Actual values of B scale calibration test blocks (HRBa) versus experimental values.*

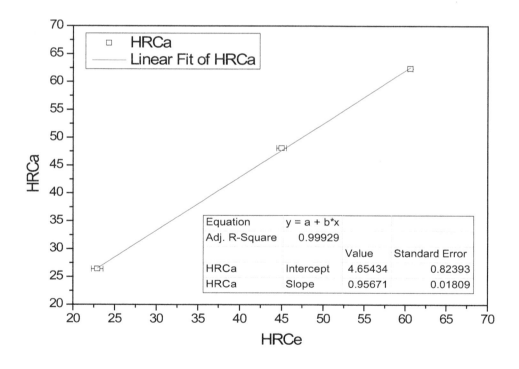

Fig. *2.4. Actual values of C scale calibration test blocks (HRCa) versus experimental values.*

Table 2.4. Scaled hardness Rockwell B and C of different steel specimens.

Sample	Scale	Scaled Rockwell Hardness Value	Standard Deviation	Tensile Strength (MPa)	HB (kg$_f$/mm²)	HB (MPa)	HV (kg$_f$/mm²)	HV (MPa)	Yield Strength (MPa)
Steel 1018 HR NHT	B	78.4	1.7	475	144	1412	144	1412	471
Steel 1018 HR HT	B	108.6	1.6						
Steel 1018 HT	C	33.3	2.4	1048	311	3050	327	3207	1069
Steel 1045 HR NHT	B	92.5	0.7	650	200	1961	200	1961	654
Steel 1045 HR NHT	C	11.5	0.9						
Steel 1045 HR HT	C	53.5	2	2010	543	5325	577	5658	1886
Steel 1045 CF NHT	B	102.3	0.5						
Steel 1045 CF NHT	C	23.2	1	810	243	2383	254	2491	830
Steel 1045 CF HT	C	55.6	2.1	2160	577	5658	613	6011	2004

deformation. The carbon content also influences the microstructure and mechanical properties of the martensite formed during the heat treatment, which plays an important role in the enhancement of the hardness and tensile strength of the steel. As the carbon content rises, the metal becomes harder and stronger, but also less ductile and more difficult to weld. A metal can be hardened by heating into the austenitic crystal phase and then cooling rapidly (quenching) to transform the high-temperature austenite (γ) phase to the hard and brittle martensite structure.

2.6. REFERENCES

[1] ASTM E18: *Standard Test Method for Rockwell Hardness and Rockwell Superficial Hardness of Metallic Materials*, ASTM International, West Conshohocken, PA, 2004.

[2] ASTM E1842: *Standard Test Method for Macro-Rockwell Hardness Testing of Metallic Materials*, ASTM International, West Conshohocken, PA, 2004.

[3] ASTM E140: *Standard Hardness Conversion Tables for Metals Relationship Among Brinell Hardness, Vickers Hardness, Rockwell Hardness, Superficial Hardness, Knoop Hardness, and Scleroscope Hardness*, ASTM International, West Conshohocken, PA, 2004.

[4] ASTM A370–07: *Standard Test Methods and Definitions for Mechanical Testing of Steel Products*, ASTM International, West Conshohocken, PA, 2004.

[5] ASM Handbook, *Mechanical Testing and Evaluation*, Vol. 8, ASM International, Materials Park, OH, 2000.

[6] *Fundamentals of Rockwell Hardness Testing*, Wilson Instruments, Division of Instron Corporation, Norwood, MA, 2004.

2.7. ASSIGNMENT

The main purpose of this lab assignment is to learn how microstructure changes can affect the indentation hardness of materials (in this case, low-carbon (AISI 1018) and medium-carbon (AISI 1045) steel).

Each student will be given six calibration blocks and six steel specimens for hardness testing, and each group will make three hardness measurements for each sample and record the results. The Rockwell hardness scale and type of indenter used should be recorded and also the minimum thickness should be checked (see table in Appendix 2A) to ensure that your sample is sufficiently thick for the scale and the indenter selected for hardness measurement.

First, perform the following tasks:

1. Scale the data as close as possible to the correct numbers corresponding to the actual hardness numbers.
2. Determine the mean and standard deviation values for each specimen.
3. Convert the Rockwell hardness to Brinell and Vicker hardness (Appendix 2B).
4. Compare the results of the different samples.

Second, discuss, analyze, and provide explanations to the following:

5. Differences in hardness before and after heat treatment; provide explanations supported by phase diagrams and TTT diagrams for each type of steel.
6. Approximate estimates of the yield strength and the ultimate tensile strength using the measured indentation hardness.
7. Indicate what you would have done differently to further increase the hardness and provide justification.

Minimum Thickness for Scale Hardness Selection Using a Diamond Indenter

| Minimum Thickness | | Rockwell Scale | | |
| | | A | | C |
in.	mm	Hardness Reading	Approximate Hardness C-Scale[1]	Dial Reading
0.014	0.36	…	…	…
0.016	0.41	86	69	…
0.018	0.46	84	65	…
0.020	0.51	82	61.5	…
0.022	0.56	79	56	69
0.024	0.61	76	50	67
0.026	0.66	71	41	65
0.028	0.71	67	32	62
0.030	0.76	60	19	57
0.032	0.81	…	…	52
0.034	0.86	…	…	45
0.036	0.91	…	…	37
0.038	0.96	…	…	28
0.040	1.02	…	…	20

Minimum Thickness for Scale Hardness Selection Using a 1/16 in (1.588 mm) Spherical Diameter Indenter

| Minimum Thickness | | Rockwell Scale | | |
| | | F | | B |
in.	mm	Hardness Reading	Approximate Hardness B-Scale[1]	Hardness Reading
0.022	0.56	…	…	…
0.024	0.61	98	72	94
0.026	0.66	91	60	87
0.028	0.71	85	49	80
0.030	0.76	77	35	71
0.032	0.81	69	21	62
0.034	0.86	…	…	52
0.036	0.91	…	…	40
0.038	0.96	…	…	28
0.040	1.02	…	…	…

[1]This table gives the approximate interrelationships of hardness values and approximate tensile strength of steels. It is possible that steels of various compositions and processing histories will deviate in hardness-tensile strength relationship from the data presented in this table. The data in this table should not be used for austenitic stainless steels, but have been shown to be applicable for ferritic and martensitic stainless steels. The data in this table should not be used to establish a relationship between hardness values and tensile strength of hard drawn wire. Where more precise conversions are required, they should be developed specially for each steel composition, heat treatment, and part.

Approximate Hardness Conversion for Non-austenitic Steels[1]
(Conversion of Rockwell C to other hardness numbers)

Rockwell C Scale, 150-kg$_f$ Load, Diamond Indenter	Vickers Hardness Number	Brinell Hardness, 3000-kg$_f$ Load, 10-mm Ball	Knoop Hardness, 500-g$_f$ Load and Over	Rockwell A Scale, 60-kg$_f$ Load, Diamond Indenter	Rockwell Superficial Hardness			Approximate Tensile Strength, ksi (MPa)
					15N Scale, 15-kg$_f$ Load, Diamond Indenter	30N Scale 30-kg$_f$ Load, Diamond Indenter	45N Scale, 45-kg$_f$ Load, Diamond Indenter	
68	940	...	920	85.6	93.2	84.4	75.4	...
67	900	...	895	85.0	92.9	83.6	74.2	...
66	865	...	870	84.5	92.5	82.8	73.3	...
65	832	739	846	83.9	92.2	81.9	72.0	...
64	800	722	822	83.4	91.8	81.1	71.0	...
63	772	706	799	82.8	91.4	80.1	69.9	...
62	746	688	776	82.3	91.1	79.3	68.8	...
61	720	670	754	81.8	90.7	78.4	67.7	...
60	697	654	732	81.2	90.2	77.5	66.6	...
59	674	634	710	80.7	89.8	76.6	65.5	351 (2420)
58	653	615	690	80.1	89.3	75.7	64.3	338 (2330)
57	633	595	670	79.6	88.9	74.8	63.2	325 (2240)
56	613	577	650	79.0	88.3	73.9	62.0	313 (2160)
55	595	560	630	78.5	87.9	73.0	60.9	301 (2070)
54	577	543	612	78.0	87.4	72.0	59.8	292 (2010)
53	560	525	594	77.4	86.9	71.2	58.6	283 (1950)
52	544	512	576	76.8	86.4	70.2	57.4	273 (1880)
51	528	496	558	76.3	85.9	69.4	56.1	264 (1820)
50	513	482	542	75.9	85.5	68.5	55.0	255 (1760)
49	498	468	526	75.2	85.0	67.6	53.8	246 (1700)
48	484	455	510	74.7	84.5	66.7	52.5	238 (1640)
47	471	442	495	74.1	83.9	65.8	51.4	229 (1580)
46	458	432	480	73.6	83.5	64.8	50.3	221 (1520)
45	446	421	466	73.1	83.0	64.0	49.0	215 (1480)
44	434	409	452	72.5	82.5	63.1	47.8	208 (1430)
43	423	400	438	72.0	82.0	62.2	46.7	201 (1390)
42	412	390	426	71.5	81.5	61.3	45.5	194 (1340)
41	402	381	414	70.9	80.9	60.4	44.3	188(1300)
40	392	371	402	70.4	80.4	59.5	43.1	182(1250)
39	382	362	391	69.9	79.9	58.6	41.9	177(1220)
38	372	353	380	69.4	79.4	57.7	40.8	171 (1180)
37	363	344	370	68.9	78.8	56.8	39.6	166(1140)
36	354	336	360	68.4	78.3	55.9	38.4	161 (1110)
35	345	327	351	67.9	77.7	55.0	37.2	156(1080)
34	336	319	342	67.4	77.2	54.2	36.1	152(1050)
33	327	311	334	66.8	76.6	53.3	34.9	149(1030)
32	318	301	326	66.3	76.1	52.1	33.7	146(1010)
31	310	294	318	65.8	75.6	51.3	32.5	141 (970)
30	302	286	311	65.3	75.0	50.4	31.3	138 (950)
29	294	279	304	64.6	74.5	49.5	30.1	135 (930)
28	286	271	297	64.3	73.9	48.6	28.9	131 (900)
27	279	264	290	63.8	73.3	47.7	27.8	128 (880)
26	272	258	284	63.3	72.8	46.8	26.7	125 (860)
25	266	253	278	62.8	72.2	45.9	25.5	123 (850)
24	260	247	272	62.4	71.6	45.0	24.3	119 (820)
23	254	243	266	62.0	71.0	44.0	23.1	117 (810)
22	248	237	261	61.5	70.5	43.2	22.0	115 (790)
21	243	231	256	61.0	69.9	42.3	20.7	112 (770)
20	238	226	251	60.5	69.4	41.5	19.6	110 (760)

[1]This table gives the approximate interrelationships of hardness values and approximate tensile strength of steels. It is possible that steels of various compositions and processing histories will deviate in hardness-tensile strength relationship from the data presented in this table. The data in this table should not be used for austenitic stainless steels, but have been shown to be applicable for ferritic and martensitic stainless steels. The data in this table should not be used to establish a relationship between hardness values and tensile strength of hard drawn wire. Where more precise conversions are required, they should be developed specially for each steel composition, heat treatment, and part.

Approximate Hardness Conversion for Non-austenitic Steels
(Conversion of Rockwell B to other hardness numbers)

Rockwell B Scale, 100-kg$_f$ Load, 1/16-in. (1.588-mm) Ball	Vickers Hardness Number	Brinell Hardness, 3000-kg$_f$ Load, 10-mm Ball	Knoop Hardness, 500-g$_f$ Load and Over	Rockwell A Scale, 60-kg$_f$ Load, Diamond Indenter	Rockwell Superficial Hardness			Approximate Tensile Strength ksi (MPa)
					Rockwell F Scale, 60-kg$_f$ Load, 1/16-in. (1.588-mm) Ball	15T Scale, 15-kg$_f$ Load, 1/16-in. (1.588-mm) Ball	30T Scale, 30-kg$_f$ Load, 1/16-in. (1.588-mm) Ball	45T Scale, 45-kg$_f$ Load, 1/16-in. (1.588-mm) Ball	
100	240	240	251	61.5	...	93.1	83.1	72.9	116 (800)
99	234	234	246	60.9	...	92.8	82.5	71.9	114 (785)
98	228	228	241	60.2	...	92.5	81.8	70.9	109 (750)
97	222	222	236	59.5	...	92.1	81.1	69.9	104 (715)
96	216	216	231	58.9	...	91.8	80.4	68.9	102 (705)
95	210	210	226	58.3	...	91.5	79.8	67.9	100 (690)
94	205	205	221	57.6	...	91.2	79.1	66.9	98 (675)
93	200	200	216	57.0	...	90.8	78.4	65.9	94 (650)
92	195	195	211	56.4	...	90.5	77.8	64.8	92 (635)
91	190	190	206	55.8	...	90.2	77.1	63.8	90 (620)
90	185	185	201	55.2	...	89.9	76.4	62.8	89 (615)
89	180	180	196	54.6	...	89.5	75.8	61.8	88 (605)
88	176	176	192	54.0	...	89.2	75.1	60.8	86 (590)
87	172	172	188	53.4	...	88.9	74.4	59.8	84 (580)
86	169	169	184	52.8	...	88.6	73.8	58.8	83 (570)
85	165	165	180	52.3	...	88.2	73.1	57.8	82 (565)
84	162	162	176	51.7	...	87.9	72.4	56.8	81 (560)
83	159	159	173	51.1	...	87.6	71.8	55.8	80 (550)
82	156	156	170	50.6	...	87.3	71.1	54.8	77 (530)
81	153	153	167	50.0	...	86.9	70.4	53.8	73 (505)
80	150	150	164	49.5	...	86.6	69.7	52.8	72 (495)
79	147	147	161	48.9	...	86.3	69.1	51.8	70 (485)
78	144	144	158	48.4	...	86.0	68.4	50.8	69 (475)
77	141	141	155	47.9	...	85.6	67.7	49.8	68 (470)
76	139	139	152	47.3	...	85.3	67.1	48.8	67 (460)
75	137	137	150	46.8	99.6	85.0	66.4	47.8	66 (455)
74	135	135	147	46.3	99.1	84.7	65.7	46.8	65 (450)
73	132	132	145	45.8	98.5	84.3	65.1	45.8	64 (440)
72	130	130	143	45.3	98.0	84.0	64.4	44.8	63 (435)
71	127	127	141	44.8	97.4	83.7	63.7	43.8	62 (425)
70	125	125	139	44.3	96.8	83.4	63.1	42.8	61 (420)
69	123	123	137	43.8	96.2	83.0	62.4	41.8	60 (415)
68	121	121	135	43.3	95.6	82.7	61.7	40.8	59 (405)
67	119	119	133	42.8	95.1	82.4	61.0	39.8	58 (400)
66	117	117	131	42.3	94.5	82.1	60.4	38.7	57 (395)
65	116	116	129	41.8	93.9	81.8	59.7	37.7	56 (385)
64	114	114	127	41.4	93.4	81.4	59.0	36.7	...
63	112	112	125	40.9	92.8	81.1	58.4	35.7	...
62	110	110	124	40.4	92.2	80.8	57.7	34.7	...
61	108	108	122	40.0	91.7	80.5	57.0	33.7	...
60	107	107	120	39.5	91.1	80.1	56.4	32.7	...
59	106	106	118	39.0	90.5	79.8	55.7	31.7	...
58	104	104	117	38.6	90.0	79.5	55.0	30.7	...
57	103	103	115	38.1	89.4	79.2	54.4	29.7	...
56	101	101	114	37.7	88.8	78.8	53.7	28.7	...
55	100	100	112	37.2	88.2	78.5	53.0	27.7	...
54	...	...	111	36.8	87.7	78.2	52.4	26.7	...
53	...	...	110	36.3	87.1	77.9	51.7	25.7	...
52	...	...	109	35.9	86.5	77.5	51.0	24.7	...
51	...	...	108	35.5	86.0	77.2	50.3	23.7	...
50	...	...	107	35.0	85.4	76.9	49.7	22.7	...
49	...	...	106	34.6	84.8	76.6	49.0	21.7	...
48	...	...	105	34.1	84.3	76.2	48.3	20.7	...

| Rockwell B Scale, 100-kg$_f$ Load, 1/16-in. (1.588-mm) Ball | Vickers Hardness Number | Brinell Hardness, 3000-kg$_f$ Load, 10-mm Ball | Knoop Hardness, 500-g$_f$ Load and Over | Rockwell A Scale, 60-kg$_f$ Load, Diamond Indenter | Rockwell F Scale, 60-kg$_f$ Load, 1/16-in. (1.588-mm) Ball | Rockwell Superficial Hardness | | | Approximate Tensile Strength ksi (MPa) |
						15T Scale, 15-kg$_f$ Load, 1/16-in. (1.588-mm) Ball	30T Scale, 30-kg$_f$ Load, 1/16-in. (1.588-mm) Ball	45T Scale, 45-kg$_f$ Load, 1/16-in. (1.588-mm) Ball	
47	...	...	104	33.7	83.7	75.9	47.7	19.7	...
46	...	...	103	33.3	83.1	75.6	47.0	18.7	...
45	...	...	102	32.9	82.6	75.3	46.3	17.7	...
44	...	...	101	32.4	82.0	74.9	45.7	16.7	...
43	...	...	100	32.0	81.4	74.6	45.0	15.7	...
42	...	...	99	31.6	80.8	74.3	44.3	14.7	...
41	...	...	98	31.2	80.3	74.0	43.7	13.6	...
40	...	...	97	30.7	79.7	73.6	43.0	12.6	...
39	...	...	96	30.3	79.1	73.3	42.3	11.6	...
38	...	...	95	29.9	78.6	73.0	41.6	10.6	...
37	...	...	94	29.5	78.0	72.7	41.0	9.6	...
36	...	...	93	29.1	77.4	72.3	40.3	8.6	...
35	...	...	92	28.7	76.9	72.0	39.6	7.6	...
34	...	...	91	28.2	76.3	71.7	39.0	6.6	...
33	...	...	90	27.8	75.7	71.4	38.3	5.6	...

Chapter 3: Monotonic Tensile Loading

Tensile testing is a standard method of determining important mechanical properties of engineering materials. In tensile testing, the axial tensile load is increased gradually in a monotonic fashion until specimen failure. American Society for Testing and Materials (ASTM) specifies the standard sample dimensions for a monotonic tensile test. During testing, the load applied to the specimen is measured by a load cell while the displacement of the frame is recorded continuously. An extensometer is often used to measure the elongation of the specimen (or change in gauge length). A load-elongation plot is generated as the test progresses from the load and elongation transducer signals, and the data are stored in a computer for subsequent analysis. Some monotonic tensile tests are performed for strength measurement only, in which case only the maximum load at the instant of fracture is of interest. Other monotonic tensile tests are performed to measure the yield strength S_Y, elastic modulus E (the slope of the linear portion of the stress-strain curve), ultimate tensile strength S_u (or strain at necking ε_u), strain at fracture ε_f (or the percent in cross-section area reduction at fracture a).

The output of a standard tensile test is a load versus displacement response characteristic of the material's response to monotonic (quasistatic) tensile loading. However, load-displacement characteristics show a dependence on specimen size. Therefore, to avoid the specimen dimension effect, a stress-strain response must be obtained from the load-elongation data to reveal the material behavior.

For uniaxial loading, the *engineering stress σ* is defined as

$$\sigma = \frac{P}{A_0} \tag{3.1}$$

where P is the instantaneous axial load and A_0 is the initial cross-section area of the specimen. The *engineering strain ε* is given by

$$\varepsilon = \frac{\Delta l}{l_0} = \frac{l - l_0}{l_0} \tag{3.2}$$

where Δl is the change in gauge length, and l_0 and l are the initial and current gauge lengths of the specimen, respectively.

Figure 3.1 shows a schematic of the σ–ε response of a specimen extended by an axial force P. Characteristic points of the σ–ε response are numbered as 1–7. A linear elastic response occurs as the load is increased from point 1 to point 2. The slope of this segment is equal to the elastic modulus E. A deviation from the linear σ–ε response is encountered with the increase of the load beyond point 2, resulting in a nonlinear elastic-plastic σ–ε response from point 3 to point 4. The critical load demarcating the onset of nonlinear (inelastic) behavior P_Y and Eq. (3.1) can be used to obtain the tensile yield strength of the material $S_Y = P_Y/A_0$. Unloading from point 4 to point 5 produces a linear elastic response and irreversible (plastic) strain ε_p (point 5). Thus, beyond the yield strength point, the total strain is

$$\varepsilon = \varepsilon_e + \varepsilon_p \tag{3.3}$$

where ε_e is the elastic strain component. Reloading from point 5 follows the path 5→4→6→7. The maximum load (stress) sustained by the material is that of point 6 and is therefore referred to as the ultimate tensile strength of the material S_u. The impression that the material undergoes softening upon a further increase of the load (indicated by the negative slope of the σ–ε response) is due to the spontaneous decrease in cross-section area at point 6, a phenomenon known as necking. The significant change in cross-section area, particularly in the case of low-strength/ductile materials, necessitates introducing the true stress and strain.

The *true strain* $\tilde{\varepsilon}$ is defined as

$$\tilde{\varepsilon} = \int_{l_0}^{l} \frac{dl}{l} = \ln\left(\frac{l}{l_0}\right) \tag{3.4}$$

From Eqs. (3.2) and (3.4), it follows that

$$\tilde{\varepsilon} = \ln(1+\varepsilon) \tag{3.5}$$

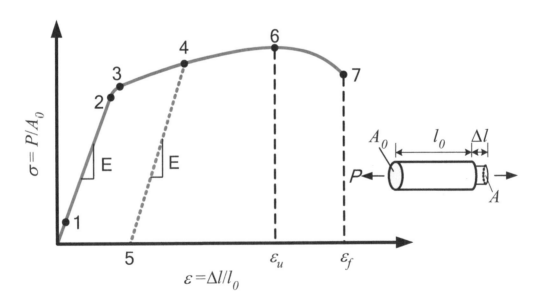

Fig. *3.1. Schematic stress-strain response of a ductile material.*

Thus, the true strain can be determined from the engineering strain calculated from the extensometer measurements as a function of applied load (engineering stress).

The *true stress* $\tilde{\sigma}$ is defined as

$$\tilde{\sigma} = \frac{P}{A} \tag{3.6}$$

from the yield point through failure as long as the amount of plastic deformation is large compared to the amount of elastic deformation.

where A is the current cross-section area of the specimen. The true stress can be expressed in terms of the engineering stress, assuming volume constancy over the gauge length $(A \cdot l = A_0 \cdot l_0)$. This assumption is valid for ~~materials that exhibit negligible necking or up to the instant of necking for relatively low-strength materials undergoing significant necking~~. Hence, using Eqs. (3.1), (3.2), and (3.6) we obtain

$$\tilde{\sigma} = \frac{P}{A} = \left(\frac{P}{A_0}\right)\left(\frac{A_0}{A}\right) = \sigma\left(\frac{l}{l_0}\right) = \sigma(1+\varepsilon) \tag{3.7}$$

After introducing the concepts of engineering and true stress and strain, we can proceed with the definition of other material properties that can be obtained from uniaxial tension loading.

Proportional limit is defined as the maximum stress at which stress and strain remain directly proportional. This proportionality is known as Hooke's law, and is expressed as $\sigma = E\varepsilon$. The proportional limit is determined from the stress-strain diagram by drawing a straight line tangent to the curve at the origin and noting the first deviation of the curve from linearity. In Fig. 3.1, the stress at point 2 is the proportional limit. Since the proportional limit depends on the measurement precision of the instrument used, it is not widely used in engineering calculations.

Poisson's ratio (v) ($0 < v < 0.5$) is the material property that relates the deformation occurring perpendicular to the direction of loading to the compressibility of the material. If the strain in the load direction is denoted by ε_{xx} and the strain in the orthogonal direction by ε_{yy}, then

$$v = -\varepsilon_{yy} / \varepsilon_{xx} \tag{3.8}$$

Elastic limit is defined as the maximum stress that the material can withstand without undergoing permanent (plastic) deformation (point 3 in Fig. 3.1). An exact determination of the elastic limit requires loading to successively higher stresses followed by unloading and gauge-length measurement to detect permanent deformation of the unloaded specimen. However, the accuracy of this procedure also depends on the instrument precision like the proportional limit. Since it is difficult to determine the elastic limit exactly by experiments, its engineering usefulness is limited.

Tensile yield strength (S_Y) (or yield point) of a material is defined as the stress at which a material begins to deform plastically. Therefore, yield strength, proportional limit, and elastic limit are equivalent parameters. When $\sigma < S_Y$, the material demonstrates purely elastic deformation, assuming its original configuration upon full unloading (spring analog). When $\sigma > S_Y$, a fraction of the deformation is permanently dissipated by the material as plastic deformation, resulting in changes in specimen length and cross-section area. The extent of these changes depends on the material behavior itself (e.g., strain hardening) and magnitude of $\sigma - S_Y$.

Thus, when $\sigma > S_Y$, the stress-strain response becomes nonlinear, following the power-law relationship

$$\tilde{\sigma} = K\tilde{\varepsilon}^n \tag{3.9}$$

where K is the *monotonic strength coefficient* and n is the *strain hardening exponent*, which is approximately equal to $\tilde{\varepsilon}_u$ (or $\tilde{\varepsilon}_p$ at necking).

Offset yield strength ($S_{Y0.2\%}$) is often obtained from the stress-strain response to quantify the yield strength of metals and alloys. $S_{Y0.2\%}$ can be determined as the intersection of the σ–ε response and a line of slope equal to E passing through the 0.2% strain point, although other strain points may be used depending on the material and application. Although this method allows for a consistent comparison of the yield strength of materials, it requires knowledge of the elastic modulus to determine the slope and the intercept with the stress-strain curve. Graphical methods always introduce errors, despite the fact that electrical and mechanical devices (e.g., extensometers) attached to the specimen aid the strain measurement. Moreover, some materials do not exhibit a true modulus (or linear response) near the origin of the stress-strain curve.

Proof stress (σ_p) is the stress at which a small amount of irreversible deformation (typically 0.1–0.2%) is obtained after full unloading.

Ultimate tensile strength (S_u) is the highest stress in the engineering σ–ε response, and is usually obtained at the instant of necking. It is calculated by dividing the maximum load sustained by the specimen by the original cross-section area, i.e., $S_u = P_{max} / A_0$. When a specimen is loaded up to (or slightly above) its ultimate tensile strength, the cross-section area exhibits a spontaneous localized decrease, resulting in the formation of a neck, where plasticity becomes excessive and the stress state differs significantly (more complex) from that in the unnecked portion of the specimen. Because of the rapid decrease in the neck cross-section area, the localized stress increases rapidly, leading to the fracture of the specimen.

Fracture strength (S_f) (or rupture strength) is referred to as the stress at the instant of fracture (rupture) of the specimen. S_f appears to be less than S_u because of the significant reduction in cross-section area after necking of ductile materials. The engineering strain at fracture ε_f and the true strength and strain at fracture, $\tilde{\sigma}_f$ and $\tilde{\varepsilon}_f$, respectively, are given by

$$\varepsilon_f = \frac{l_f - l_0}{l_0} \ , \ \tilde{\sigma}_f = \frac{P_f}{A_f} \ , \ \tilde{\varepsilon}_f = \ln\left(\frac{A_0}{A_f}\right) \tag{3.10}$$

where subscript f indicates that the particular parameter is evaluated at fracture.

Percent elongation (%EL) refers to the elongation at fracture (rupture), is indicative of the ductility of the material, and can be obtained as

$$\%EL = \left(\frac{l_0 - l_f}{l_0}\right) \times 100 \tag{3.11}$$

Percent reduction in area (%RA) refers to the reduction in cross-section area of the specimen at the instant of fracture (rupture), is also characteristic of the ductility of the material, and is given by

$$\%RA = \left(\frac{A_0 - A_f}{A_0}\right) \times 100 \tag{3.12}$$

It is noted that $\tilde{\varepsilon}_f$, %EL, and %RA can be used to evaluate the ductility (or toughness) of the material. Ductility is a property describing the extent to which materials can be deformed plastically without fracture. Another means of describing the ductility of the tested material is by calculating the area under the $\tilde{\sigma}$–$\tilde{\varepsilon}$ curve, defined as the specific strain energy density u_p given by

$$u_f = \int_0^{\tilde{\varepsilon}_f} \tilde{\sigma} d\tilde{\varepsilon} = \int_0^{\varepsilon_Y} E\varepsilon d\varepsilon + \int_{\varepsilon_Y}^{\tilde{\varepsilon}_f} K\tilde{\varepsilon}^n d\tilde{\varepsilon} = \frac{S_Y \varepsilon_Y}{2} + \frac{K\left(\varepsilon_f^{n+1} - \varepsilon_Y^{n+1}\right)}{n+1} \tag{3.13}$$

For a high-toughness material, $\tilde{\varepsilon}_f \gg \varepsilon_Y$, and using Eq. (3.9), we can simplify Eq. (3.13) to the following:

$$u_f \approx \frac{K\varepsilon_f^{n+1}}{n+1} = \frac{\tilde{\sigma}_f \tilde{\varepsilon}_f}{n+1} \tag{3.14}$$

Equation (3.14) shows an inverse relationship between toughness and strain hardening.

3.2. TEST EQUIPMENT

Tensile testing is usually performed with a high-load capacity Instron machine. A 150-kN-force Instron machine will be used in this experiment. The load is applied by hydraulic means using pressurized oil. For safety and to protect the machine, the maximum load applied to the specimen should be less than 100 kN.

The machine is controlled by a computer workstation and can be operated in three different modes: load, position, and strain mode, implying control of either load, extension, or engineering strain during testing. Figure 3.2 shows photographs of the entire Instron machine and important components, such as safety grip, fixture, control panel, emergency stop button, extensometer, and load cell.

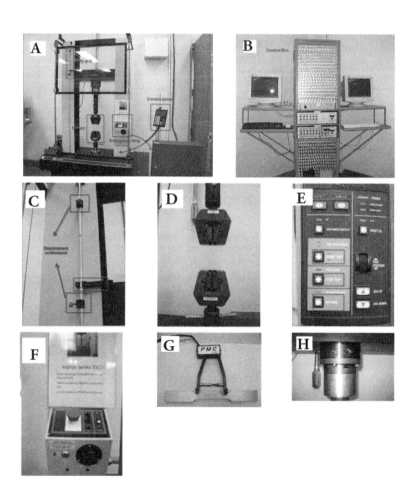

Fig. *3.2. Photographs of tensile testing machine and important components: (A) Instron machine, (B) control system, (C) safety grip, (D) fixture, (E) control panel, (F) emergency stop, (G) extensometer, and (H) load cell.*

The specimen dimensions for a standard tensile test are defined by ASTM code B557–06 [1]. The specimen may have rectangular (Fig. 3.3) or circular (Fig. 3.4) cross-section area. In the present experiment we will use aluminum and steel specimens with rectangular cross-section areas, such as that shown in Fig. 3.5.

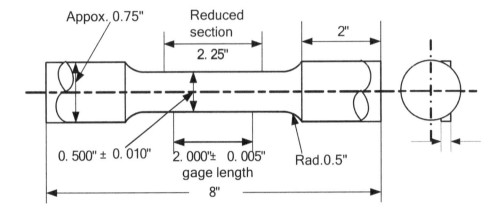

Fig. *3.3. Tensile specimen with rectangular cross-section area and dimensions as specified by the ASTM code [1].*

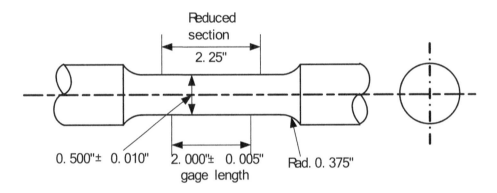

Fig. *3.4. Tensile specimen with circular cross-section area and dimensions as specified by the ASTM code [1].*

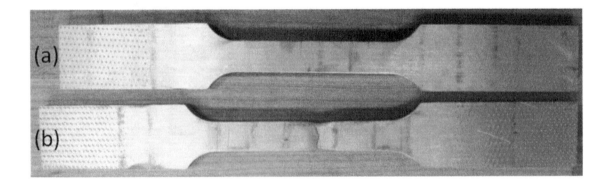

Fig. *3.5. Standard ASTM tensile specimen of 6061 aluminum alloy with rectangular cross-section area before (a) and after (b) fracture by tensile loading.*

1. Measure and record the initial dimensions of the tensile test specimen. Although all specimens are supposed to have the same dimensions as the standard specimen, there may be some tolerance variations.
2. Turn on the Instron machine, if not already turned on, and allow for a few minutes to stabilize.
3. Open the Instron software termed "Merlin" in the control computer, go to "File -> Open method" and load the tensile test program "ME108 Tensile Test.mtM".
4. Set all the parameters required by the software, including specimen dimensions, loading speed (elongation (strain) rate), etc. When finished, go to "OK -> Run method" to return to the data acquisition interface. Settings can be checked in the "Test control".
5. Press the "Return" button on the Instron control panel next to the machine. This will bring the fixture to the original position.
6. Load the tensile specimen and clamp it tightly with the fixture.
7. Zero the load and the strain on the machine control panel or software interface.
8. Press "Start" on the control panel or software to start the tensile test.
9. Monitor the load and strain data output. If there is anything unusual, press the "Stop" button on the control panel.
10. Upon specimen fracture, the Instron machine will stop by itself.
11. Save the raw data file into your own folder through "File -> Data -> End & Save".
12. Carefully remove the fractured specimen.

3.5. REPRESENTATIVE RESULTS

Standard tensile specimens made of 6061 aluminum (Al) alloy and AISI 1045 steel were used to perform monotonic tensile tests up to fracture. Output files included raw data of force and displacement. True stress and strain were calculated from the recorded engineering stress and strain raw data using the formulas given in Sect. 3.1. The monotonic loading tests are summarized in Table 3.1.

Table 3.1. Materials and elongation (strain) rates used in the monotonic loading tests.

Test number	Material	Elongation rate (mm/min)	Extensometer
1	Al 6061 Al alloy	3	No
2	6061 Al alloy	3	Yes
3	6061 Al alloy	6	No
4	6061 Al alloy	6	Yes
5	AISI 1045 steel	6	No
6	AISI 1045 steel	6	Yes

Figure 3.6 shows *engineering* and *true* stress-strain responses of 6061 Al alloy for an elongation rate of 3 and 6 mm/min. A comparison of Figs. 3.6(a) and 3.6(b) shows that increasing the elongation rate by a factor of two did not yield discernible differences in the mechanical response, suggesting that 6061 Al alloy does not exhibit strain rate sensitivity in this range of elongation rate. These stress-strain responses can be used to extract information for the elastic and plastic material properties described in section 3.1. For example, by extracting the data corresponding to the proportional part of the stress-strain curve, the elastic modulus can be determined as the slope of a linear fit through the raw data, as demonstrated in Fig. 3.7 for the stress-strain response of Fig. 3.6(a).

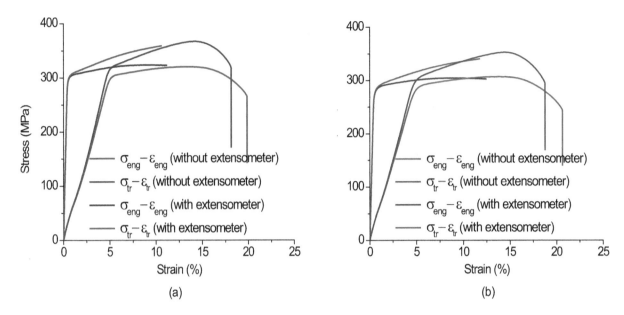

Fig. *3.6. True stress (σ_r) versus true strain (ε_{tr}) and engineering stress (σ_{eng}) versus engineering train (ε_{eng}) responses of 6061 Al alloy for an elongation rate equal to (a) 3 and (b) 6 mm/min obtained with and without the use of an extensometer.*

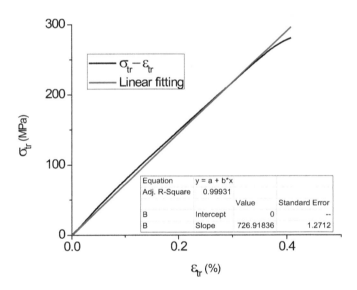

Fig. *3.7. Linear elastic regime of true stress (σ_{tr}) versus true strain (ε_r) response of 6061 Al alloy.*

Figure 3.8 shows true and engineering stress-strain responses of AISI 1045 steel for an elongation rate of 6 mm/min obtained with and without the use of an extensometer. The stress-strain curves of Fig. 3.8(a) show the material response up to the instant of fracture (indicated by a sharp stress decrease). Although the use of the extensometer improves the accuracy of the strain measurements, it prevents loading of the material all the way to fracture because this can damage the extensometer. In contrast to the 6061 Al alloy, this steel material exhibits yielding phenomena, demonstrated by the irregularities at the yield point.

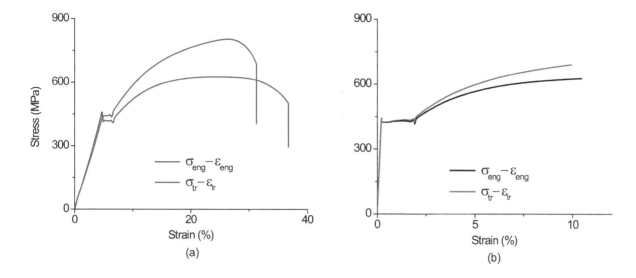

Fig. *3.8. Comparison of true stress (σ_{tr}) versus true strain (ε_{tr}) and engineering stress (σ_{eng}) versus engineering train (ε_{eng}) plots of AISI 1045 steel for an elongation rate equal to 6 mm/min: (a) stress-strain curves obtained without the extensometer for a tensile load up to fracture and (b) stress-strain curves obtained with the extensometer up to a strain level of ~10%.*

Table 3.2 gives the elastic modulus, yield point (stress and strain), ultimate tensile strength, and fracture strength of 6061 Al alloy and AISI 1045 steel obtained from the stress-strain responses shown in Figs. 3.6 and 3.8.

Table 3.2. Mechanical properties of 6061 Al alloy and AISI 1045 steel extracted from monotonic loading tests.

Material (elongation rate)	Elastic modulus (GPa)	Yield Point		Ultimate tensile strength (MPa)	Fracture strength (MPa)
		Strain (%)	Stress (MPa)		
6061 Al alloy (3 mm/min)	72.7	0.47	293	367	324
6061 Al alloy (6 mm/min)	65.3	0.47	270	353	296
AISI 1045 steel (6 mm/min)	220.3	0.22	427	804	686

Should be taken from stress–strain curve

3.6. REFERENCES

[1] ASTM Standard B557: Nonferrous Metal Standards and Nonferrous Alloy Standards: Standard Test Methods for Tension Testing Wrought and Cast Aluminum- and Magnesium-Alloy Products, ASTM International, West Conshohocken, PA (DOI: 10.1520/B0557-06).

3.7. ASSIGNMENT

The purpose of this lab assignment is to provide insight into the measurement of the mechanical properties of materials from monotonic loading (tensile) tests using dog-bone shape 6061 Al alloy and AISI 1045 steel specimens.

Obtain and discuss the following:

1. Plot the *engineering* and *true* stress-strain responses of each material on the same figure and discuss similarities and differences between the two types of responses.

2. Identify on each plot of 6061 Al alloy the following material properties and show calculations (where needed): (a) elastic modulus, (b) yield strength (check if there is an upper and a lower yield stress), (c) ultimate tensile strength, (d) strength and strain at fracture, (e) total strain at necking, (f) strain energy at fracture, and (g) percent reduction in area both at necking and fracture.

3. Compare the stress-strain responses of 6061 Al alloy and AISI 1045 steel and discuss the underlying reasons for any differences that you observe.

Chapter 4: Fracture Toughness

4.1. BACKGROUND

Fracture toughness can be defined as the energy needed to fracture a specimen of the material of interest. The fracture energy can be obtained from a standard impact test (known as the Charpy test) involving the impact of a V-notched specimen by a swing pendulum of known mass (Fig. 4.1). The main task is to determine the minimum energy to fracture the specimen due to impact of a swing pendulum of a given mass.

The test is based on an energy balance principle. The fracture energy E_f (i.e., the energy absorbed by the fractured specimen) is proportional to the loss of kinetic energy of the pendulum; hence, $E_f \propto (h - h')$, where h and h' are the initial and final (after impact) height of the specimen, respectively.

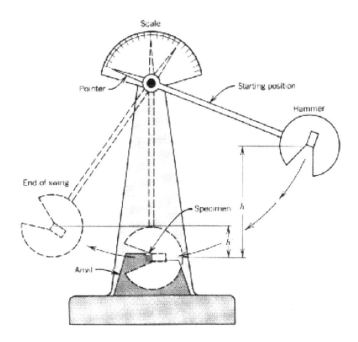

Fig. *4.1. Schematic of the Charpy impact testing apparatus [1].*

The impact tester is a Sonntag universal impact machine (model SI-1, serial no. 492253). Figure 4.2(a) shows the impact Charpy tester and Fig. 4.2(b) shows the front panel of the tester. Table 4.1 gives the available initial energy (potential) of the tester.

Table 4.1. Initial energy of Charpy impact tester.

Pendulum		Initial energy (lbf·ft)	
Light head	No weight	60	25
	Added weight	120	50
Heavy head		240	100

(a)   (b)

Fig. *4.2. (a) Standard Charpy impact test machine and (b) front panel of impact tester.*

4.3. TEST SPECIMEN

Standard specimens consisting of AISI 1045 steel will be used to perform fracture toughness experiments. Some of the specimens will be first heat treated at 850°C for 1 h and then quenched to 170°C in 3 s to obtain microstructures significantly different from the as-received steel microstructure. Figure 4.3 shows the dimensions of the fracture toughness specimen, as specified by the ASTM A370 code [2].

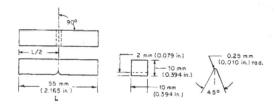

Note 1—Permissible variations shall be as follows:

Notch length to edge	90 ±2°
Adjacent sides shall be at	90° ± 10 min
Cross-section dimensions	±0.075 mm (±0.003 in.)
Length of specimen (L)	+ 0, − 2.5 mm (+ 0, − 0.100 in.)
Centering of notch (L/2)	±1 mm (±0.039 in.)
Angle of notch	±1°
Radius of notch	±0.025 mm (±0.001 in.)
Notch depth	±0.025 mm (±0.001 in.)
Finish requirements	2 µm (63 µin.) on notched surface and opposite face; 4 µm (125 µin.) on other two surfaces

Fig. *4.3. Dimensions of standard specimen used in the Charpy impact test [2].*

The main steps of the experimental protocol for impact testing are:
- (a) Raise the pendulum to the latched position.
- (b) Position and center the specimen on the supports against the anvils.
- (c) Release the pendulum smoothly. The specimen will be impacted by the striker.
- (d) After impact, record the impact energy absorbed by the specimen to the nearest $lb_f \cdot ft$.
- (e) Recover the pieces of the fractured specimen, observe them with a microscope, and identify the fracture behavior (brittle versus ductile).
- (f) Lock the pendulum in its original position for the next impact test.
- (g) Some additional important details:
    - The impact machine should be calibrated to zero absorbed energy before mounting the specimen.
    - For safety, a block should be inserted between the pendulum and the specimen base after fracturing the specimen.
    - Each specimen must be centered on the specimen base before each test.
    - The rough faces of the heat-treated samples should be polished off before testing.
    - To prevent recording an erroneous value caused by jarring the indicator when locking the pendulum in its upright (ready) position, the value for each test should be read from the indicator prior to locking the pendulum for the next test.
    - In selecting a scale for the initial potential energy it should be considered that absorbed energy >80% of the scale range is inaccurate.
    - The axial position of the anvil must be fixed by tightening the pendulum bearing.

(a)   (b)

Fig. *4.4. (a) Specimen supports and (b) pendulum bearing of Charpy impact test machine.*

4.5. REPRESENTATIVE RESULTS

Representative results are presented in this section for untreated and heat-treated AISI 1040 steel. In all tests, the initial pendulum energy was equal to 100 $lb_f \cdot ft$. Table 4.2 shows the effect of heat treatment on the impact (adsorbed) energy (toughness) of the steel specimens. The hardness of the untreated and heat-treated steel specimens is also listed for comparison.

Table 4.2. Heat treatment effect on impact energy and hardness of AISI 1040 steel specimens.

Sample condition		Impact energy ($lb_f \cdot ft$)							Rockwell hardness C
		1	2	3	4	5	Average	Standard deviation	Average [3]
Hot-rolled	nHT	32	36	34	36	32	34	2	7.15
	HT	48	38	43	44*	22	39	10.15	51.09

*invalid test (not considered)

nHT = non heat-treated; HT = heat-treated

Table 4.3. Impact energy of AISI 1040 steel [4].

Condition	As-rolled	Normalized	Annealed
Impact energy ($lb_f \cdot ft$)	36	48	32.7

(a)   (b)   (c)

Fig. *4.5. Fracture surfaces of (a) heat-treated and (b) untreated AISI 1040 steel and (c) schematic of a fracture surface characterized by ductile behavior [5].*

The slightly higher impact (fracture) energy of the heat-treated specimens than that of the untreated specimens indicates that fracture toughness was not significantly affected by the heat treatment, although the impact energy data of the heat-treated specimens yielded a higher standard deviation. Considering the impact energy data of AISI 1040 steel given in the literature (Table 4.3), the average impact energy of the heat-treated specimens (39 $lb_f \cdot ft$) is close to that of the as-rolled steel (36 $lb_f \cdot ft$). The data given in Table 4.2 reveal that the heat treatment process increased significantly the hardness and appreciably the fracture toughness of the steel specimens. Thus, both the resistance against plastic deformation (hardness) and the resistance against cracking (toughness) were enhanced as a result of heat treatment.

The cross-section surface of the heat-treated specimens (Fig. 4.5(a)) was smooth in most areas, implying a brittle behavior (Fig. 4.5(c)), while the cross-section of the untreated specimens (Fig. 4.5(b)) demonstrated ductile behavior characterized by all-around dull shear regions.

4.6. REFERENCES

[1] Callister, W. D., *Materials Science and Engineering: An Introduction*, John Wiley, New York, 2003, p. 208.

[2] ASTM Standard: *A 370 Standard Test Methods and Definitions for Mechanical Testing of Steel Products*, 2007.

[3] Data provided by S.G. McCormick.

[4] *ASM Metals Reference Book*, American Society for Metals, Metals Park, OH, 1983, p. 211.

[5] ASTM Standard: *E 23 Standard Test Methods for Notched Bar Impact Testing of Metallic Materials*, 2006.

4.7. ASSIGNMENT

The purpose of this lab is to provide insight into the measurement of the fracture toughness of materials using impact load testing. Impact experiments will be performed with V-notched AISI 1045 steel specimens, both untreated and heat-treated.

Because of the limited time and relatively large number of specimens required for legitimate statistical analysis of the fracture toughness data, each group will test >10 specimens of either untreated or heat-treated AISI 1045 steel and will use the data of the other group performing fracture toughness tests at the same time in the lab report.

[handwritten: 5 specimens from 2 sample groups 1 specimen from 3rd group (+ others)]

The following information should be provided:

1. A table including the impact energy of each fractured specimen (indicate which specimens were measured by you and which by the other group) and calculated mean and standard deviation values (e.g., see Table 4.2).

2. Representative cross-section photographs of the fractured untreated and heat-treated steel specimens (each report should contain photos of both untreated and heat-treated steels).

3. A discussion of the effect of heat treatment on fracture toughness and the type of fracture behavior exhibited by the untreated and heat-treated steel specimens.

Chapter 5: Time- and Rate-Dependent Deformation

5.1. BACKGROUND

Polymers consist of long-chain molecules formed primarily by carbon-carbon bonds. The simplest example is polyethylene ($-(CH_2)_n-$), which is obtained from the polymerization reaction of ethylene (C_2H_4). In general, polymers possess relatively low density, low strength, and time-dependent mechanical behavior (viscoelasticity).

Polymers used as engineering materials can be classified into three main groups: thermoplastics, thermosetting plastics, and elastomers.

Thermoplastics (e.g., polyethylene, polystyrene, etc.) are generally composed of long coiling chains which are not crosslinked, and are classified as low-strength, low-modulus polymers. They tend to soften and often melt when heated, and obtain their original solid condition when cooled down.

Thermosetting Polymers (epoxies, unsaturated polyesters, etc.) usually consist of chains that are crosslinked to form three-dimensional networks. Chain crosslinking enhances polymer strength upon curing and prevents softening upon reheating.

Elastomers (natural rubber, polychloroprene, etc.) are distinguished from other plastics by their rubbery behavior. They can be deformed at significantly large strains (e.g., 200% or even higher), with most deformation being recovered after the removal of stress. Long-chain elastomers do not crosslink and generally return to their coiled configuration (original shape) when unloaded.

5.2. MECHANICAL BEHAVIOR OF POLYMERS

The tests used to evaluate the mechanical properties of polymers are similar to those used for metals and alloys. However, although stress, strain, elastic modulus, strength, creep, fracture, etc., have the same meaning for polymers as for metals and alloys, polymers exhibit viscoelastic behavior, which is strongly dependent on temperature and strain rate. Most linear-chain polymers are relatively soft and ductile, whereas complex-network polymers are generally relatively hard and brittle.

Figure 5.1(a) shows typical stress-strain responses of polymers exhibiting brittle (curve A), plastic (curve B), and highly elastic behavior (curve C). Usually, the stress-strain response is highly dependent on temperature and strain rate, as seen in Fig. 5.1(b). The effect of increasing strain rate is equivalent to decreasing the temperature.

Figure 5.2 shows that the elastic behavior of polymers is highly dependent on temperature; for example, the elastic modulus of a thermoplastic polymer varies with temperature. There are four distinct regions of viscoelastic behavior. At relatively high temperatures, thermoplastics melt, becoming like viscous liquids, while at relatively low temperatures, thermoplastics behave as rigid solids, demonstrating elastic deformation. The transition from elastic-to viscous-dominated mechanical behavior is demarcated by the glass transition temperature T_g. Semicrystalline and crystalline polymers demonstrate a sudden change in specific volume due to the collapse of the crystalline structure when the temperature exceeds the melting temperature T_m, as seen in Fig. 5.2(b).

For practical applications, elastomers are used at temperatures well above their T_g to retain the molecular mobility necessary for elastic behavior. Thermoplastics are generally used at temperatures above T_g because of their higher toughness and ductility at temperatures above T_g. Thermosets are often used at temperatures below T_g in order to take advantage of their structural rigidity.

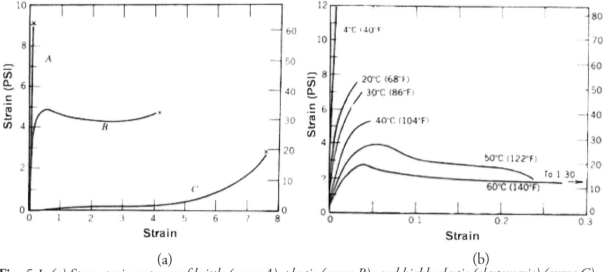

Fig. 5.1. (a) Stress-strain response of brittle (curve A), plastic (curve B), and highly elastic (elastomeric) (curve C) polymers [1] and (b) dependence of polymer stress-strain response on temperature [2].

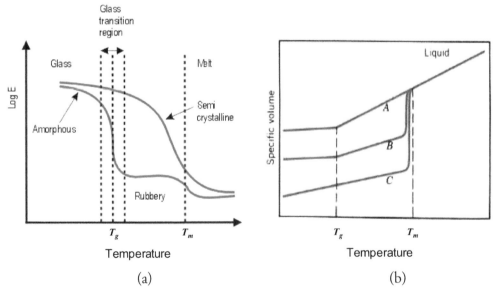

Fig. 5.2. (a) Temperature effect on elastic modulus and (b) transition temperature of thermoplastic polymers with amorphous (curve A), semicrystalline (curve B), and crystalline (curve C) microstructures [2,3].

Many semicrystalline polymers exhibit a unique deformation behavior, such as that shown in Fig. 5.3. Usually, at the start of necking, thinning of the cross-section area spreads along the entire specimen length. This phenomenon is a result of uncoiling, alignment, and stretching of the molecular chains. Further deformation leads to chain alignment resulting in the formation of crystalline lamellae, breaking of large lamellae into smaller segments, and, finally, breaking of the bonds between molecular chains.

Creep and stress relaxation can occur as a result of the relatively gradual sliding of polymer chains with respect to each other during loading. Creep often occurs in polymers exhibiting continued deformation under constant stress. Stress relaxation refers to the decrease in stress required to maintain a specific strain. Stress relaxation is attributed to the internal structure reorganization resulting from the movement of polymer chains, the breaking and reforming of secondary bonds between the chains, and the mechanical untangling and recoiling of the chains. Stress relaxation in polymers is highly dependent on temperature and/or strain rate.

Stress relaxation can be described by a generalized Maxwell model (Fig. 5.4), represented by Eq. (5.1), which can be used to fit the stress relaxation data.

$$\sigma = \sigma_0 + \sigma_1 \exp\left(-t/\tau_1\right) + \sigma_2 \exp\left(-t/\tau_2\right) + \ldots \qquad (5.1)$$

where σ is the stress after time t and σ_0 is the steady-state stress.

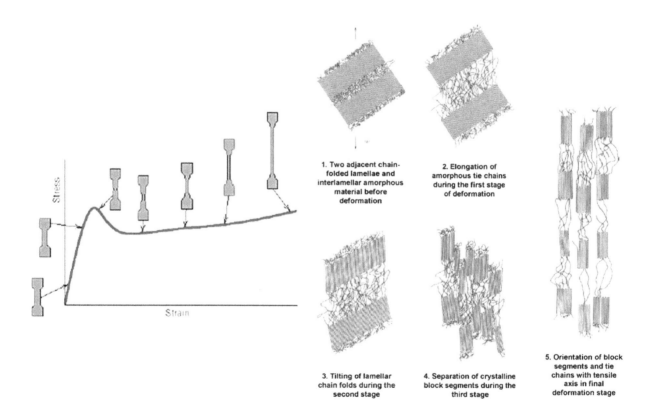

Fig. *5.3. Schematic of tensile stress-strain curve of a semicrystalline polymer at room temperature and corresponding changes in internal structure [4].*

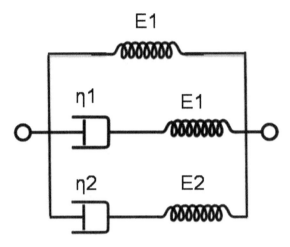

Fig. *5.4. Maxwell model characterized by a second-order exponential decay function.*

5.3. EXPERIMENTAL PROCEDURES

5.3.1. Material

Specimens consisting of UHMWPE (purchased from McMaster-Carr) with geometry and dimensions in accord with the ASTM code (i.e., width = 0.5 in., thickness = 0.25 in., and length = 2.5 in.) will be used to perform these experiments.

5.3.2. Equipment

Figure 5.5 shows an Instron machine which will be used to evaluate the behavior of UHMWPE. Eye protection is always required when operating the machine. To operate the machine, a key is used to turn on part 1, which can be obtained from the machine shop staff. Then, the software on the computer on the right side is opened. It will take a few minutes to establish communication between the machine and the computer. When the computer is ready, mount the polymer sample on the machine, then load the program for the specific experiment on the computer and set the appropriate program parameters. Before starting the experiment, make sure to set the proper locations of the pair of stop pins in part 2, the machine will stop automatically if one of the pins is reached to protect the machine. Therefore, the upper pin should be high enough for the sample to fail, and usually for UHMWPE it should be much higher than other experiments since the sample can elongate up to more than 300% before it fractures. The distance between the moving part and the lower pin should be less than the distance between the two holder heads to prevent crashing into each other if something goes wrong. After that, press 1 and 2 on the control panel (part 3) to zero the strain and load, and then press "Start" to start the experiment. "Stop" can be used to stop the machine if there is a problem. If "Reset GL" is pressed before a specific experiment, "Return" can be used to bring the holder heads to the original place to load the new sample after finishing the previous experiment. All the settings on the control panel can be completed using the computer software interface. The red button on the blue box near part 3 is the emergency stop.

5.3.3. Uniaxial tensile loading experiment

1. Use the Instron sample holder for tensile testing, turn on the machine with the key, and log in both computers (should be done by GSI or machine shop staff).
2. Load specimen and tight the two holders.
3. Open Instron software *Merlin* and select the polymer tensile test programs. If the current program is not the right one, go to "File -> Open method" to load the right program.
4. Confirm the tensile test parameters: "Edit method -> Setup", check the sample parameters (e.g., dimensions) and test parameters. Click "OK -> Run method" to return to data acquisition interface. Settings can also be checked in "Test control".
5. Zero the load and strain on the machine control panel or on the software interface.
6. Press "Start" on the control panel or in the software to start the experiment.
7. Observe the load versus strain curve; watch the deformation of the sample.
8. Wait until the sample fails due to rupture, save the acquired data, "File -> Data -> End & save", specify the sample number in the name, click "OK".
9. Unload the failed sample and mark the sample number; record corresponding data.
10. Repeat the experiment with a new sample using a different strain rate by loading new programs.

5.3.4. Stress relaxation experiment

1. Load a new specimen.
2. Change previous program to polymer stress relaxation programs.
3. Confirm test parameters: "Edit method -> Setup". Check sample parameters (e.g., dimensions) and test parameters (e.g., loading speed). Click "OK -> Run method" to return to data acquisition interface.
4. Open "Test control -> Event", check that the actions like "Hold extension" for condition "Load =1.5 kN" are in "Enable", and close the window.
5. Zero load and extension on Instron control panel or in software.
6. Press "Start" on the control panel or in software to start the experiment.
7. Observe the stress relaxation of the specimen shown by the stress-time response.
8. Wait until data acquisition has been completed.
9. Save the acquired data, unload the specimen, mark the specimen number, and record corresponding data.
10. Repeat the stress relaxation experiment with a new sample using different conditions by loading new programs.

Fig. *5.5. Equipment for polymer tensile testing.*

5.4. REPRESENTATIVE RESULTS

5.4.1. Uniaxial deformation of UHMWPE at different strain rates

Figure 5.6 shows stress vs. strain (engineering) responses of UHMWPE deformed at different strain rates $\dot{\varepsilon}$ (or extension speed). For an extension speed of 5 mm/min, specimen fracture (rupture) did not occur after 20 min of testing and, therefore, the experiment was stopped. However, for an extension speed of 15, 50, and 500 mm/min, rupture occurred after about 15, 4, 0.35 min, respectively, as demonstrated by the sharp stress drops in Fig. 5.6. The stress-strain curves show a significant strain rate effect on the mechanical behavior of UHMWPE. The higher the strain rate, the more pronounced the strain hardening effect and the smaller the rupture strain.

Figure 5.7 shows photographs of ruptured UHMWPE specimens. Significant necking was not observed, and all the specimens were uniformly stretched until rupture. Specimen curving after rupture is due to the release of a large residual stress, especially in the case of high rupture strain.

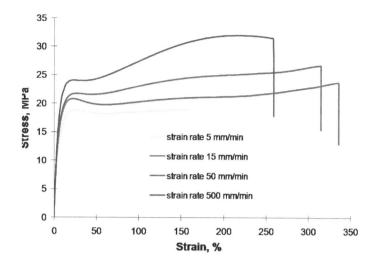

Fig. *5.6. Stress versus strain response of UHMWPE stretched at different strain rates.*

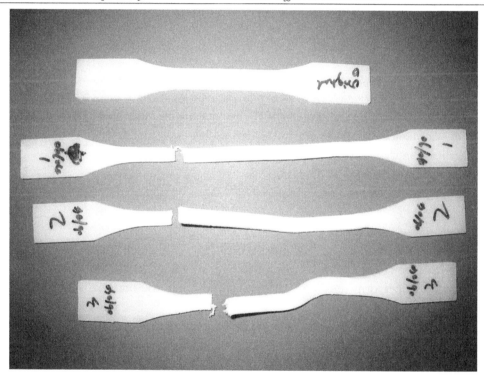

Fig. *5.7. UHMWPE tensile specimens (from top to bottom: original specimen and ruptured specimens for extension speed equal to 15, 50, and 500 mm/min).*

5.4.2. Stress relaxation of UHMWPE

Figure 5.8 shows representative stress relaxation results for UHMWPE stretched under different conditions. For extension speed equal to 50 and 250 mm/min, the specimen was first loaded up to 1.5 kN and then held at a constant elongation for 10 min to observe the relaxation of the stress. It can be seen that the stress decreased much faster in the case of higher preload strain rate. In the case of the specimen pre-strained by 100% and then held at constant elongation, the starting stress is higher. The three responses shown in Fig. 5.8 can be curve fitted by Eq. (5.1) (second-order form) to extract the unknown material parameters.

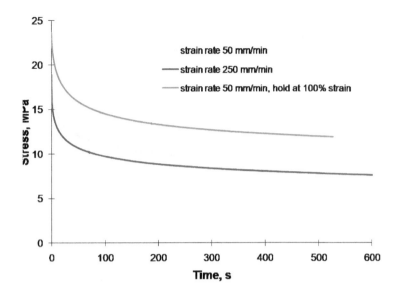

Fig. *5.8. Stress relaxation of UHMWPE after stretching under different conditions.*

5.5. REFERENCES

[1] Callister, W. D., *Fundamentals of Materials Science and Engineering*, John Wiley, 2001.

[2] Carswell, T. S., and Nason, H. K., "Effect of Environmental Conditions on the Mechanical Properties of Organic Plastics," *Symposium on Plastics*, American Society for Testing and Materials, 1944.

[3] www.azom.com/details.asp?articleID=83

[4] Schultz, J. M., *Polymer Materials Science*, Prentice-Hall, 1974.

[5] Dowling, N. E., *Mechanical Behavior of Materials*, Prentice-Hall, Englewood Cliffs, NJ, 2007.

5.6. ASSIGNMENT

The objective of this lab is to study the mechanical behavior of viscoelastic materials like polymers, using a thermoplastic polymer to perform tensile experiments, and relate the material stress-strain response to the microstructure characteristics.

You are asked to perform tensile tests and stress relaxation experiments of a representative thermoplastic polymer, ultrahigh molecular weight polyethylene (UHMWPE), and study the stress-strain response under different testing conditions, e.g., strain rate.

Obtain and discuss the following results:

1. Perform monotonic stretching experiments with UHMWPE specimens at strain rates in the range of 0.24–7.87 min^{-1} (or extension speeds of 15–500 mm/min) and stress relaxation experiments (constant elongation) after deforming the specimens at relatively low and high strain rates corresponding to extension speeds of 50 and 250 mm/min, respectively.

2. Plot the *engineering* stress-strain response at different testing conditions and discuss differences in the deformation behavior and failure characteristics (e.g., elongation at fracture, necking, etc.) in terms of testing conditions.

3. Plot the stress relaxation responses at different testing conditions, determine the relaxation parameters, and comment on the underlying reasons for any differences (e.g., microstructure effects, such as chain recoiling, stretching, and sliding).

4. Obtain photographs of the ruptured specimens and discuss characteristic features.

Chapter 6: Deformation due to Cyclic Loading

6.1. BACKGROUND

Components of machines, vehicles, and structures are frequently subjected to repeated loads, and the resulting cyclic stresses (strains) may trigger microscale damage processes, eventually leading to macroscopic failure, such as fracture. Even at stresses below the yield strength of the material, microscopic damage can progressively accumulate with repeated (cyclic) loading to result in void nucleation and, subsequent, crack initiation and propagation, ultimately causing catastrophic gross failure of the component known as fatigue. Although elastic and plastic material properties can be determined from monotonic tensile tests, where a standard specimen is subjected to monotonically increasing elongation up to rupture/fracture, the stress-strain response of most materials under cyclic loading differs from that under monotonic loading. Therefore, cyclic loading tests according to ASTM specifications must be performed to assess the stress-strain response of a material under cyclic stress (strain) conditions.

When a component consisting of an elastic-plastic material is subjected to cyclic loading, it may respond in four different ways [1], as shown schematically in Fig. 6.1. At sufficiently light loads, the yield condition is not satisfied at any material point and the response is *purely elastic* (Fig. 6.1(a)). However, if yielding occurs in some regions of the component during initial loading but the residual stresses produced upon unloading and/or strain hardening of the material cause a purely elastic steady-state response in subsequent cycles, the material is said to exhibit *elastic shakedown* (Fig. 6.1(b)). The maximum load for elastic shakedown is referred to as the *elastic shakedown limit*. If the elastic shakedown limit is exceeded, plastic deformation will occur in each loading cycle with two possible scenarios. If the steady-state response is characterized by a closed cycle of plastic strain (*cyclic plasticity*), it is described as a state of *plastic shakedown*. This response occurs when the load is less than the maximum load for plastic shakedown, known as the *plastic shakedown limit*. Above the plastic shakedown limit, cyclic loading leads to steady accumulation of unidirectional plastic strain, and this incremental process is referred to as *ratcheting*.

The most common cyclic test involves a complete strain (stress) reversal between two strain limits of equal magnitude (i.e., stress ratio $R = \varepsilon_{max}/\varepsilon_{min} = -1$), as illustrated in Fig. 6.2 [2]. The specimen is loaded in tension up to a maximum tensile strain $\varepsilon_{max} = \varepsilon_a$, and the loading direction is then reversed until the specimen reaches a strain $\varepsilon_{\min} = -\varepsilon_a$, where ε_a is the strain amplitude, defined as $\varepsilon_a = \Delta\varepsilon / 2 = (\varepsilon_{\max} - \varepsilon_{\min}) / 2$. This loading sequence is applied at a fixed frequency, and the stress necessary for this to occur is recorded as a function of time (middle plots in Fig. 6.2). The stresses needed to enforce the strain limits usually change as the test progresses. Recorded stress data are then cross-plotted with corresponding strain data to obtain stress-strain responses revealing strain hardening or softening of the material due to cyclic loading (top and bottom right-side plots in Fig. 6.2, respectively). If the stress increases with cycles of fixed strain range $\Delta\varepsilon$, the material is said to exhibit *cyclic hardening behavior*, whereas if the stress decreases the material exhibits *cyclic softening behavior* [3].

For most engineering metals and alloys, cyclic hardening or softening is usually rapid at first, but the change from one cycle to the next decreases with increasing number of cycles. Often, the behavior becomes approximately stable, meaning that further changes in the stress-strain response can be neglected as secondary.

If the stress-strain variation during stable behavior is plotted for a full stress-strain cycle, a closed-loop hysteresis is obtained, as shown schematically in Fig. 6.3. This stable deformation state represents *plastic*

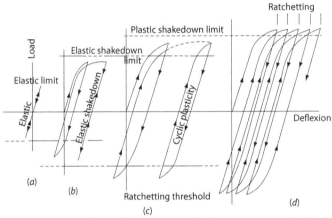

Fig. 6.1. *Schematic material response to cyclic loading: (a) purely elastic response for loads below the elastic limit; (b) plastic deformation during initial loading cycles followed by purely elastic steady-state response* (elastic shakedown); *(c) for loads above the elastic shakedown limit, steady-state deformation is characterized by a closed cycle of plastic deformation* (cyclic plasticity); *this deformation process is known as* plastic shakedown; *(d) for loads above the plastic shakedown limit* (ratcheting threshold), *the material accumulates plastic deformation in each loading cycle; this incremental plastic deformation process is referred to as* ratcheting *[1].*

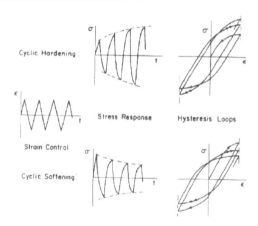

Fig. 6.2. *Cyclic strain-control test and possible stress-strain responses corresponding to cyclic hardening (top) and cyclic softening (bottom) [2].*

shakedown. Upon the reversal of the loading direction, the stress-strain path initially exhibits a constant slope close to the elastic modulus E, determined from a uniaxial tension test, and later deviates from linearity as the material undergoes plastic deformation in the reverse direction, i.e., the material undergoes sequential tension-compression loadings, resulting in stress and strain ranges of $\Delta\sigma$ and $\Delta\varepsilon$, respectively. The area of the stress-strain hysteresis loop represents the *dissipation energy* at the particular loading cycle.

Figure 6.4 shows the strain amplitude versus fatigue life with characteristic hysteresis loops shown at typical high, medium, and low strain amplitude levels. It can be seen that for high-cycle fatigue the hysteresis loop is much smaller than that for low-cycle fatigue, indicating less amount of energy dissipated irreversibly [4].

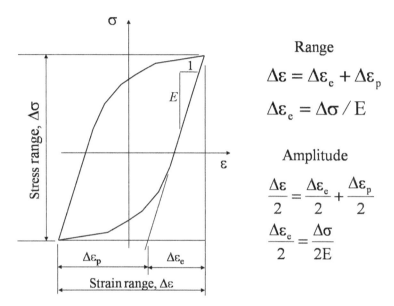

Range

$$\Delta\varepsilon = \Delta\varepsilon_e + \Delta\varepsilon_p$$

$$\Delta\varepsilon_e = \Delta\sigma / E$$

Amplitude

$$\frac{\Delta\varepsilon}{2} = \frac{\Delta\varepsilon_e}{2} + \frac{\Delta\varepsilon_p}{2}$$

$$\frac{\Delta\varepsilon_e}{2} = \frac{\Delta\sigma}{2E}$$

Fig. *6.3. Stable stress-strain hysteresis loop (plastic shakedown).*

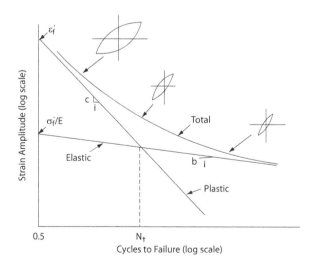

Fig. *6.4. Stable hysteresis loops and cyclic stress-strain curve constructed from corresponding amplitudes of stress and strain of different hysteresis loops.*

6.2.1. Test Specimen

Figure 6.5 shows the standard specimen for tensile testing specified by ASTM code A 370–07 [5]. The experiments will be performed with 0.5-in.-wide sheet-type specimens with dimensions in accord with the former ASTM code. Only the round corner design of the polycarbonate specimen is slightly different from the standard specimen.

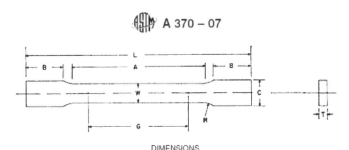

ASTM A 370 – 07

DIMENSIONS

	Standard Specimens				Subsize Specimen	
	Plate-Type, 1½-in. Wide		Sheet-Type, ½-in. Wide		¼-in. Wide	
	in.	mm	in.	mm	in.	mm
G—Gage length (Notes 1 and 2)	8.00 ± 0.01	200 ± 0.25	2.000 ± 0.005	50.0 ± 0.10	1.000 ± 0.003	25.0 ± 0.08
W—Width (Notes 3, 5, and 6)	1½ + ⅛ − ¼	40 + 3 − 6	0.500 ± 0.010	12.5 ± 0.25	0.250 ± 0.002	6.25 ± 0.05
T—Thickness (Note 7)			Thickness of Material			
R—Radius of fillet, min (Note 4)	½	13	½	13	¼	6
L—Over-all length, min (Notes 2 and 8)	18	450	8	200	4	100
A—Length of reduced section, min	9	225	2¼	60	1¼	32
B—Length of grip section, min (Note 9)	3	75	2	50	1¼	32
C—Width of grip section, approximate (Notes 4, 10, and 11)	2	50	¾	20	⅜	10

Fig. *6.5. ASTM A 370–07 standard of specimen dimensions for tensile testing.*

6.2.2. Equipment

Cyclic loading tests will be carried out with an Instron series 5500 equipment of maximum tensile load equal to 80 kN (Fig. 6.6).

(a)   (b)   (c)

Fig. *6.6. Instron equipment for fatigue testing: (a) Instron tensile tester, (b) computer unit for data collection and post-processing, and (c) control panel.*

6.2.3. Testing Procedure

The following main steps must be followed in the cyclic loading tests:

1. Install the tensile test sample and mount the extensometer at the center part (gauge length) of the fixed sample carefully (the extensometer is fragile and expensive).
2. Open the programmed method (test recipe) and check the parameters to ensure consistency with the test you are going to perform.
3. Calibrate the extensometer in the software, zero the extension, and balance the load.
4. Start the planned test. (Note: Observe the force-extension curve displayed using the software and stop the test manually if the force or extension exceeds the set maximum limit).
5. Save the data after completing the test.
6. Zero the load (actually set to a small value, e.g., ~0.1 kN) by adjusting the position knob on the control panel.
7. Remove the sample by gently hammering it and return the grips to the initial position by pressing the return button on the control panel.

6.2.4. Designed Experiments

Cyclic tensile testing includes four strain-controlled tests and two load-controlled tests. To avoid a discontinuity in the stress-strain curve when the loading direction is reversed, all tests are designed to be in the tensile range, with the following test parameters (where F = force, ε = strain, $\Delta\varepsilon$ = strain range, N = cycles (reversals)):

1. ε_{max} = 0.0325, $\quad \varepsilon_{min}$ = 0.03, $\quad \Delta\varepsilon$ = 0.0025, $\quad \bar{\varepsilon}$ = 0.03125, $\quad N$ = 25
2. ε_{max} = 0.0685, $\quad \varepsilon_{min}$ = 0.0675, $\quad \Delta\varepsilon$ = 0.001, $\quad \bar{\varepsilon}$ = 0.068, $\quad N$ = 25
3. ε_{max} = 0.07, $\quad \varepsilon_{min}$ = 0.0675, $\quad \Delta\varepsilon$ = 0.0025, $\quad \bar{\varepsilon}$ = 0.06875, $\quad N$ = 25
4. ε_{max} = 0.0715, $\quad \varepsilon_{min}$ = 0.0675, $\quad \Delta\varepsilon$ = 0.004, $\quad \bar{\varepsilon}$ = 0.0695, $\quad N$ = 25
5. F_{max} = 50 N, $\quad F_{min}$ = 40 N, $\quad \Delta F$ = 10 N, $\quad \bar{F}$ = 45 N, $\quad N$ = 25
6. F_{max} = 50 N, $\quad F_{min}$ = 20 N, $\quad \Delta F$ = 10 N, $\quad \bar{F}$ = 35 N, $\quad N$ = 25

6.3. REPRESENTATIVE RESULTS

The objective of the results presented in this section is to provide an idea of the expected stress-strain responses and a format of how to present the collected data. However, due to sample-to-sample differences, the test results shown here may differ from those in your tests. Figures 6.7–6.14 show representative results from cyclic tests performed with 1045 steel specimens. Results from both displacement-control and load-control experiments are discussed.

Figures 6.7 and 6.8 show that cyclic loading under strain control in the range of 3%–3.25% resulted in elastic shakedown. Figure 6.7 shows the entire cyclic loading procedure, including initial elastic deformation, yielding, post-yield strain hardening, and cyclic loading behaviors. As shown in Fig. 6.8, after the first partial unloading, subsequent cycles are essentially elastic because the loading and unloading curves overlap, indicating that energy dissipation due to plastic deformation did not occur. High-cycle fatigue would be expected for components subjected to this kind of cyclic loading.

Figures 6.9–6.11 show that cyclic loading under strain control in the range of 6.75%–7.0% resulted in plastic shakedown. Figure 6.9 shows the load versus time, whereas Fig. 6.10 shows the entire cyclic loading procedure. The magnified region of the cyclic response shown in Fig. 6.11 reveals that the loading and the unloading paths of the cyclic response do not overlap. The produced stable hysteresis imply plastic

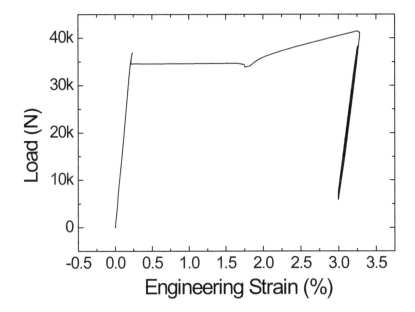

Fig. *6.7. Load versus engineering strain response of AISI 1045 steel (3%–3.25% strain control).*

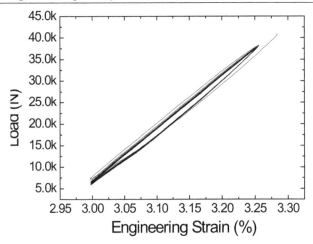

Fig. *6.8. Cyclic load versus engineering strain response of AISI 1045 steel showing the development of a hysteresis loop (3%–3.25% strain control).*

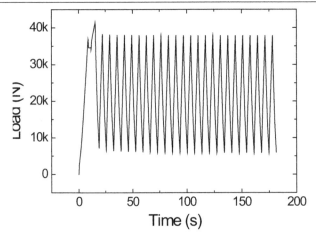

Fig. *6.9. Cyclic load versus time response of AISI 1045 steel (6.75%–7% strain control).*

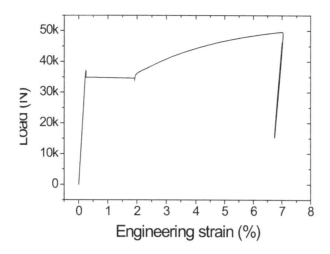

Fig. *6.10. Load versus engineering strain response of AISI 1045 steel (6.75%–7% strain control).*

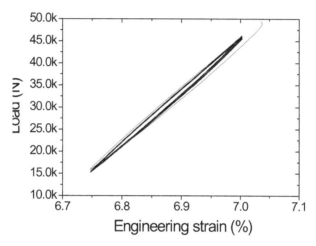

Fig. *6.11. Cyclic load versus engineering strain response of AISI 1045 steel showing the development of a hysteresis loop (6.75%-7% strain control).*

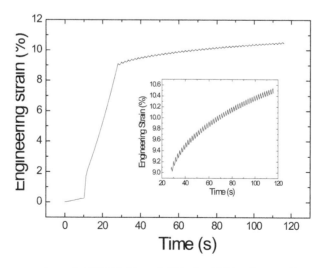

Fig. *6.12. Engineering strain versus time of AISI 1045 steel (40–50 kN load control).*

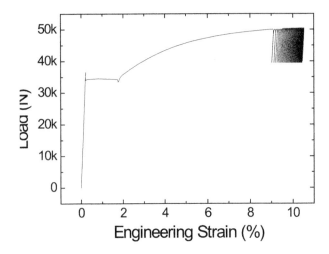

Fig. *6.13. Cyclic load versus engineering strain response of AISI 1045 steel (40–50 kN load control).*

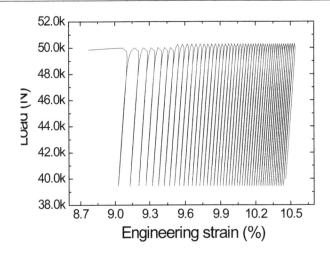

Fig. *6.14. Cyclic load versus engineering strain resposne of AISI 1045 steel demonstrating ratchettng behavior (40–50 kN load control).*

shakedown. Thus, increasing the cyclic strain range from 3%–3.25% to 6.75%–7.0% induced a change in the steady-state material response from elastic to plastic shakedown. The hysteresis area represents the amount of energy dissipated in the form of plastic deformation per cycle. Low-cycle fatigue would be expected for components subjected to this kind of cyclic loading.

Figures 6.12–6.14 show that cyclic loading in the range of 40–50 kN (load control experiments) resulted in ratcheting. Figure 6.12 shows the strain versus time, whereas Fig. 6.13 shows the entire response. Conversely to the previous cases, significant accumulation of plastic deformation occurred in this case during cyclic loading (Fig. 6.14). This progressive accumulation of plastic deformation would eventually lead to failure after applying a relatively low number of loading cycles.

6.4. REFERENCES

[1] Kapoor, A., and Johnson, K. L., "Plastic Ratcheting as a Mechanism of Metallic Wear," *Proceedings Royal Society of London, Series A*, Vol. 445, 1994, pp. 367–381.

[2] From [landgrave 70]; copyright © ASTM

[3] Dowling, N. E., *Mechanical Behavior of Materials*, 3rd ed., Prentice-Hall, Englewood Cliffs, NJ, 1998, pp. 639–654.

[4] Dowling, N. E., *Mechanical Behavior of Materials*, 3rd ed., Prentice-Hall, Englewood Cliffs, NJ, 1998, pp. 715–718.

[5] ASTM A 370–07: *Standard Test Methods and Definitions for Mechanical Testing of Steel Products*.

6.5. ASSIGNMENT

The objective of this lab assignment is to provide hands-on the mechanical behavior of 1045 steel subjected to repetitive load/unload reversals. Specifically, the effect of tensile loading applied repetitively (cyclic loading) should be correlated to the resulting hardening/softening material response applied on each specimen, and the obtained results should be analyzed in the context of cyclic hardening (softening), plastic shakedown (constant plastic strain accumulation per cycle), ratcheting (continuous accumulation of plasticity), and energy dissipated in each load/unload reversal (loading cycle).

The following tasks should be performed throughout:
1. Collection of raw data of engineering stress and engineering strain for each test.
2. Plot of engineering stress versus engineering strain for each test.
3. Determination of type of material response, i.e., elastic, elastic shakedown, plastic shakedown, ratcheting, in each test.
4. Identification of steady-state hysteresis loop, calculation of area enclosed by this loop for tests producing plastic shakedown behavior, identification of steady-state cycle of strain increment, and calculation of strain increase per cycle for tests yielding ratcheting behavior.
5. Comparison of results from tests (1) and (3) and discussion of the effect of mean strain on the material response to cyclic loading.
6. Comparison of results from tests (2), (3), and (4) and discussion of the effect of strain amplitude on the material response due to cyclic loading.
7. Comparison of results from tests (5) and (6) and discussion of the effect of upper and lower load on the material response due to cyclic loading.
8. Comparison of results from tests (5), (6), and (7) in the context of elastic-plastic deformation theory, considering the correlation between strain energy density and deformation and/or any microscopic argument to rationalize your analysis.
9. Relation of obtained results to any practical engineering phenomenon, such as fatigue and fracture, and discussion of pertinent implications.

You are asked to perform the following experiments, supplemented by analysis and discussion:
1. Cyclic tensile loading tests with 1045 steel specimens (6 tests).
2. Identify the type of material response for each test, i.e., elastic, elastic shakedown, plastic shakedown, or ratcheting.
3. Hysteresis loop (stress versus strain) plot and calculation of the energy dissipated per cycle.
4. Comparison and discussion of the differences between stress-strain hysteresis loops and the effect of strain amplitude on the resulting cyclic behavior.
5. Comparison of results from tests where the specimens demonstrated ratcheting behavior, calculation of strain increment per cycle, and discussion of the effect of load (amplitude) on ratcheting behavior.

Chapter 7: Fatigue Testing

Fatigue is a type of failure caused by cyclic loading at stress (strain) levels below those of failure under static loading. Fatigue is generally understood to be a process dominated by cyclic plastic deformation, implying that fatigue damage can occur even at stress levels below the monotonic yield strength. Fatigue damage is a cumulative process comprising, in general, three stages: (a) *crack initiation*, characterized by the formation of microcracks in regions containing stress raisers and/or defects, such as notches, nonmetallic inclusions, and preexisting crack-like defects, (b) *crack growth*, dominated by crack propagation, and (c) *fracture*, involving sudden failure caused by a dominant crack. Thus, the *fatigue life* N_f (usually expressed in terms of stress cycles or strain reversals) is the sum of the periods corresponding to the previously mentioned fatigue stages. Because the duration of the third stage (fracture) is negligibly small compared to crack initiation (N_i) and crack growth (N_g), the fatigue life can be expressed as $N_f = N_i + N_g$. The fatigue life of brittle materials is controlled by crack initiation (i.e., $N_f \approx N_i$), whereas that of ductile materials is controlled by crack growth (i.e., $N_f \approx N_g$). Thus, engineering the microstructure of brittle and ductile materials to enhance N_i and N_g, respectively, is of great practical importance.

There are three main approaches to analyze fatigue failure: (a) stress-based approach, (b) strain-based approach, and (c) fracture mechanics approach. The first two approaches are classified as *macroscopic* because N_f is directly related to a "global" driving force of fatigue (e.g., stress or strain amplitude, $\Delta\sigma$ or $\Delta\varepsilon$, respectively), whereas the third approach is classified as *microscopic* because N_f is obtained in terms of the "local" driving force of fatigue (e.g., stress intensity factor range, ΔK). The focus in this lab assignment is on macroscopic approaches to fatigue, specifically the stress-based approach.

Figure 7.1 is a schematic of a semi-log stress-life (*S-N*) plot. (An established nomenclature in the field is to denote $\Delta\sigma$ by S and N_f by N.) The *S-N* approach is normally used when the applied stress is primarily within the elastic range of the material and the resultant lives (cycles to failure) are long. The *S-N* approach does not work well in low-cycle fatigue (e.g., $N_f < 10^3$ cycles) applications where the resulting strains have a significant plastic component. The plot shown in Fig. 7.1 reveals two distinct regions of finite and infinite fatigue life, demarcated by a fatigue life on the order of ~10^6 stress cycles. In the fatigue life region ~$10^3 < N_f < $ ~10^6, the stress amplitude and the number of fatigue cycles follow the power-law relationship [1]:

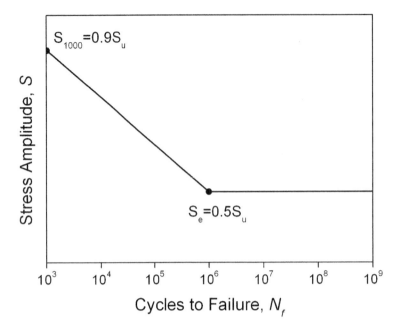

Fig. *7.1. Schematic of stress-life (S-N) curve.*

$$S = AN_f^b \tag{7.1}$$

where A and b are fatigue parameters intrinsic of the material, which can be related to the stress amplitude corresponding to $\sim 10^3$ stress cycles (S_{10^3}) and the stress amplitude resulting in infinite fatigue life (S_e), known as the *endurance limit*. Thus, Eq. (7.1) can be rewritten as

$$S = \frac{S_{10^3}^2}{S_e} \times N_f^{-\frac{1}{3}\log\left(\frac{S_{10^3}}{S_e}\right)} \tag{7.2}$$

Equation (7.2) can be simplified by using the empirical relationships $S_{10^3} \approx 0.9 S_u$ and $S_e \approx 0.5 S_u$, where S_u is the ultimate tensile strength obtained from monotonic loading tests; thus,

$$S \approx 1.62 S_u N_f^{-0.085} \tag{7.3}$$

The strain-based approach to fatigue considers both the elastic and the plastic strain components that may occur in localized regions where fatigue cracks initiate. Stresses and strains in such regions are analyzed and used as a basis for life estimates. Figure 7.2 shows a log-log plot of the strain amplitude $\Delta\varepsilon/2$ versus strain reversals $2N_f$. It is noted that the strain-life approach measures fatigue life in reversals ($2N_f$), while the stress-life method uses cycles (N); thus, one reversal is one-half of a full cycle.

The $\Delta\varepsilon/2$ versus $2N_f$ curve can be approximated by two asymptotes, referred to as plastic and elastic asymptotes because they represent the elastic and plastic components of the strain range, $\Delta\varepsilon_e$ and $\Delta\varepsilon_p$, respectively, i.e.,

$$\frac{\Delta\varepsilon}{2} = \frac{\Delta\varepsilon_e}{2} + \frac{\Delta\varepsilon_p}{2} = \frac{\sigma_f'}{E}(2N_f)^b + \varepsilon_f'(2N_f)^c \tag{7.4}$$

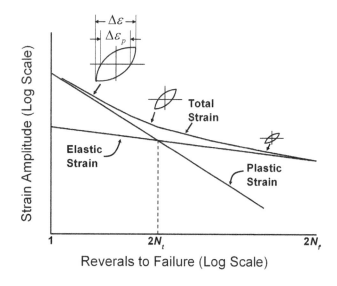

Fig. *7.2. Schematic of strain-life ($\Delta\varepsilon/2$ versus $2N_f$) curve.*

where

σ'_f = fatigue strength coefficient (the true stress at failure in one reversal)

ε'_f = fatigue ductility coefficient (the true strain at failure in one reversal)

b = fatigue ductility exponent

c = fatigue strength exponent

A prime is used to avoid confusion with similar parameters used in monotonic loading. Some fairly good approximations of the fatigue parameters are: $\sigma'_f \approx \sigma_f$ (corrected for necking) and $\varepsilon'_f \approx \varepsilon_f$.

The *transition life* $2N_t$, determined by the intersect of the two asymptotes ($\Delta\varepsilon_e = \Delta\varepsilon_p$), is given by

$$2N_t = \left(\frac{E\varepsilon'_f}{\sigma'_f}\right)^{\frac{1}{b-c}} \tag{7.5}$$

Thus, the transition life (typically, $2N_t \approx 10^6$) represents the fatigue-life at which the elastic and plastic strain ranges are equal to each other. Schematic representations of the shape of the hysteresis loop at different lives are also shown in Fig. 7.2. When $2N_f < 2N_t$, plastic strain plays a more dominant role in fatigue life (*low-cycle fatigue*) and the hysteresis loop is wide, whereas when $2N_f > 2N_t$, the fatigue life is controlled by elastic strains and the hysteresis loop area is narrow.

There are many parameters affecting the fatigue-life, such as type of loading, surface quality, material type, environment, and temperature. Figure 7.3 shows the effect of mean stress σ_0 on the strain-life curve (Morrow's correction [2]).

Morrow [2] suggested that the mean stress can be accounted for by modifying the elastic term of Eq. (7.4) as following

$$\frac{\Delta\varepsilon_e}{2} = \frac{\Delta\sigma}{2E} = \left(\frac{\sigma'_f - \sigma_0}{E}\right)(2N_f)^b \tag{7.6}$$

Substituting Eq. (7.6) in Eq. (7.4) gives

$$\frac{\Delta\varepsilon}{2} = \left(\frac{\sigma'_f - \sigma_0}{E}\right)(2N_f)^b + \varepsilon'_f(2N_f)^c \tag{7.7}$$

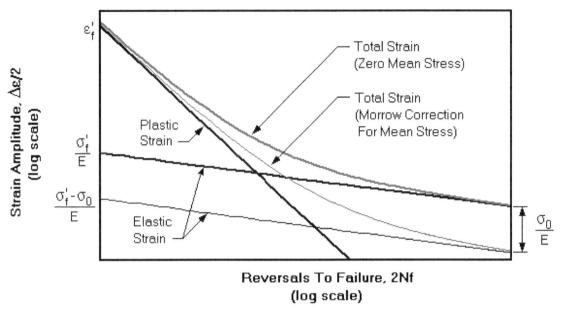

Fig. *7.3. Effect of mean stress σ_0 on strain-life curve [3].*

Manson and Halford [4] modified both the elastic and the plastic terms of Eq. (7.4) and obtained the following strain-life relationship

$$\frac{\Delta\varepsilon}{2}=\frac{\sigma'_f-\sigma_0}{E}\left(2N_f\right)^b+\varepsilon'_f\left(1-\frac{\sigma_0}{\sigma'_f}\right)^{c/b}\left(2N_f\right)^c \qquad (7.8)$$

The cyclic stress-plastic strain relationship $\sigma=K'\varepsilon_p^{n'}$, where K' and n' are the cyclic strength coefficient and the cyclic strain hardening exponent, respectively, and Massing's hypothesis [5] that the stabilized hysteresis loop may be obtained by doubling the cyclic stress-strain curve can be used to obtain the hysteresis curve equation given below

$$\Delta\varepsilon=\Delta\varepsilon_e+\Delta\varepsilon_p=\frac{\Delta\sigma}{E}+2\left(\frac{\Delta\sigma}{2K''}\right)^{1/n'} \qquad (7.9)$$

The effect of stress concentration (notch effect) is included in the fatigue life predictions through the fatigue stress concentration factor K_f, which is a function of the theoretical stress concentration factor K_t and the notch sensitivity q of the material, i.e.,

$$K_f=1+q\left(K_t-1\right),\, 0\le q\le 1 \qquad (7.10)$$

The theoretical stress concentration factor is a function of geometry, whereas the notch sensitivity varies between zero (notch insensitive material) and one (fully notch sensitive material). Figures 7.4 and 7.5 show the notch effects on K_t and q, respectively.

For different heat treatment conditions, the material would show different fatigue properties. Figure 7.6 shows strain-life curves for medium-carbon steel for two different heat treatment conditions. For a given strain, the high-strength (quenched) steel exhibits longer fatigue life in the high-cycle fatigue regime (where elastic strains are dominant) and shorter fatigue life in the low-cycle fatigue regime (where plastic strains are dominant) compared to the ductile (normalized) steel. The reason is that, as the ultimate strength of steel increases, the transition life decreases and elastic strains dominate for a greater portion of the fatigue life range.

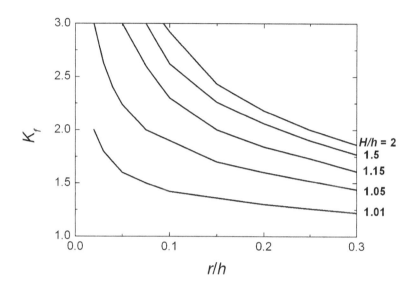

Fig. *7.4. Notch radius effect on theoretical stress concentration factor K$_t$.*

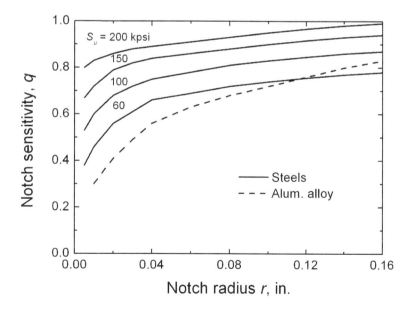

Fig. *7.5. Notch radius effect on notch sensitivity q.*

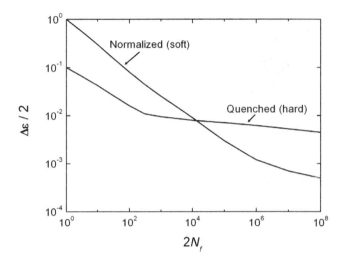

Fig. *7.6. Effect of heat treatment conditions on strain-life curve of steels.*

7.2. EXPERIMENTAL PROCEDURES

7.2.1. Test Specimen

Straight shank specimens consisting of as-received AISI 1045 steel were machined in accord to the standards set by the ISO code 1143 [6]. Figure 7.7 shows the standard dimensions of the straight shank fatigue specimen.

7.2.2. Testing Equipment

Fatigue tests will be performed with an Instron R.R. Moore high-speed rotating beam fatigue testing apparatus, shown in Fig. 7.8, with a loading capacity in the range of 9–101 lb and a maximum rotational speed of ~10,000 rpm. For safety, the machine is equipped with an automatic specimen-fail switch that immediately stops the motor as soon as the specimen fails (fractures). The straight shank specimen is mounted in two specimen holders. Loading harness and weight pan assemblies are used to apply the desired load to the specimen.

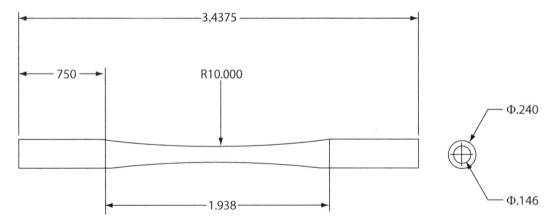

Fig. *7.7. Dimensions of standard straight shank specimen for fatigue testing*

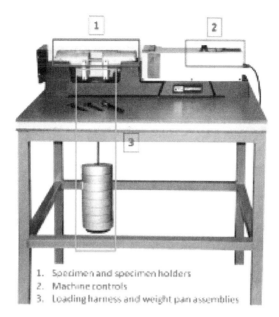

1. Specimen and specimen holders
2. Machine controls
3. Loading harness and weight pan assemblies

Fig. *7.8. Instron R.R. Moore apparatus for fatigue testing.*

Schematics of machine controls are shown in Fig. 7.9. The power entry module (1) accepts the incoming power for the machine. The *motor on* button (4) is used to start the motor, whereas the *motor off* button (3) is used to stop the motor. The rotational speed is set by the *speed control* knob (2), while the rotational (fatigue) cycles are recorded by the *revolution counter* (5).

Figure 7.10 shows schematically the loading applied to the fatigue specimen (W = total load applied to the specimen, L = moment arm (i.e., distance from the end support to the load point, fixed at 4 inches), and D = minimum diameter of cylindrical fatigue specimen). The central part of the specimen between the two load points is subjected to a constant bending moment equal to $0.5\,WL$. For any cross-section of the central part of the specimen, the upper part is subject to a compressive stress, while the bottom part is subject to a tensile stress. The maximum normal stress σ_{max} in the cylindrical specimen is given by

$$\sigma_{max} = \frac{M \times D/2}{I} = \frac{WL/2 \times D/2}{\pi D^4/64} = \frac{16WL}{\pi D^3} \qquad (7.11)$$

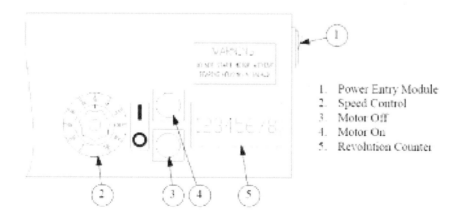

1. Power Entry Module
2. Speed Control
3. Motor Off
4. Motor On
5. Revolution Counter

Fig. *7.9. Controls of Instron R.R. Moore fatigue testing apparatus.*

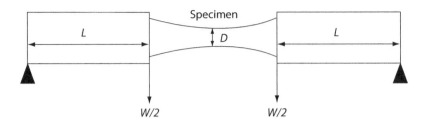

Fig. *7.10. Schematic of loading applied to the fatigue specimen.*

7.2.3. Testing Procedure

The following main steps must be followed to perform the fatigue tests.

1. Measure and record the initial dimensions of the fatigue specimen, especially the minimum diameter of the specimen.
2. Load the specimen on the machine (a technician will guide you through this process).
3. Set up the loading harness and the weight pan assemblies on the machine, starting with a minimum weight of 9 lb.
4. Set up the rotational speed to 50% (~6,000 rpm) using the *speed control* knob. (Actually, the speed control knob gives a percentage of the maximum output voltage of the motor control.)
5. Turn on the *power entry module*.
6. Press the *motor on* button to start the motor. (The specimen must be mounted onto the tester before the motor can be operated to prevent damage of the coupling.)
7. After the drive motor reaches a steady-state speed (about several hundreds of cycles), gently load the additional weight desired for the particular experiment on the weight pan and reset the *revolution counter*.
8. Once the specimen has failed (fractured), record the number of cycles (reversals) shown by the *revolution counter*.
9. Turn off the *power entry module*, unload the loading harness and the weight pan assemblies, and, finally, dismount the failed specimen.
10. Repeat steps 2–10 for each fatigue specimen for different testing conditions.

7.3. REPRESENTATIVE RESULTS

7.3.1. Fatigue tests of notched specimens

Figure 7.11 shows the fatigue life N_f as a function of stress amplitude S of as-received AISI 1045 steel. It can be seen that the higher the stress amplitude, the shorter the fatigue life. Crack growth rate depends on the stress intensity factor range ΔK (driving force). Since the increase in S intensifies ΔK, faster crack growth and, in turn, shorter fatigue life was obtained at higher stress amplitudes.

As mentioned earlier, in the finite fatigue life region ($\sim 10^3 < N_f < \sim 10^6$) the stress amplitude and the fatigue cycles follow a power-law relationship. This power-law relationship for AISI 1045 steel, obtained by curve-fitting the experimental data shown in Fig. 7.11, is given by

$$S = 1.93 \times 10^2 \, N_f^{-0.08} \, (\text{ksi}) \tag{7.12}$$

Based on this curve fitting, the endurance limit S_e is found equal to ~63 ksi. Since the ultimate tensile strength S_u is about two times the S_e, it follows that $S_u \approx 126$ ksi (= 860 MPa), which is close to $S_u = 804$

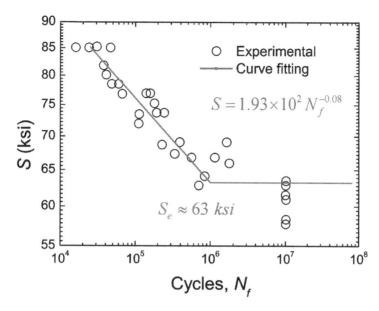

Fig. *7.11. Stress amplitude S vs. fatigue life (cycles) N_f for AISI 1045 steel.*

MPa of AISI 1045 steel measured from monotonic tensile tests (Table 3.2). Appendix A gives estimates of the time to failure (or cycles to failure) as a function of applied load.

7.3.2. Optical microscopy examination of fatigued specimens

A comparison between experimental data and estimated results of fatigue life is given in Table 7.1 for different stress amplitudes. The fatigue lives estimated from Eq. (7.12) differ from the fatigue lives determined experimentally because Eq. (7.12) is a statistical result while the specimens used for fatigue testing exhibit variations, such as differences in defect density, dimensional tolerances, etc., resulting in different fatigue lives under the same fatigue loading conditions.

Table 7.1. *Experimental and calculated results of fatigue life at different stress amplitudes.*

Load W (lb)	Moment arm L (in.)	Specimen minimum diameter D (in.)	Stress amplitude S (ksi)	Fatigue life N_f (cycles)	
				Calculated [Eq. (7.12)]	Experimental
10.0	4	0.147	64.13	957,637	843,063
11.5	4	0.146	75.28	129,113	179,351
13.0	4	0.146	85.10	27,883	23,891

Figures 7.12–7.14 show optical microscopy photographs of fracture cross-sections of specimens fatigued at different stress amplitudes. The significantly rougher cross-sections obtained at relatively high stress amplitudes are attributed to faster crack growth and, possibly, unstable crack propagation, especially at the later stage of cracking. The fact that cross-section edges are generally rougher suggests that fracture

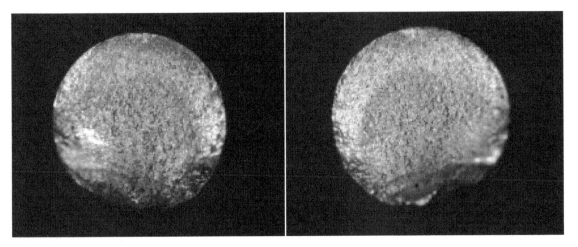

Fig. *7.12. Optical microscope photographs of specimens fatigued under S = 64.13 ksi.*

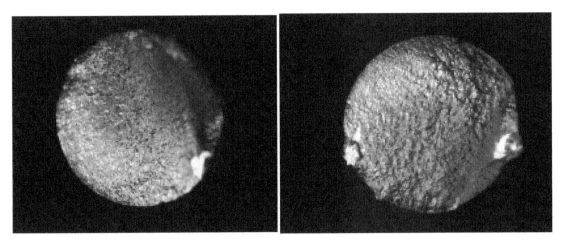

Fig. *7.13. Optical microscope photographs of specimens fatigued under S = 75.28 ksi.*

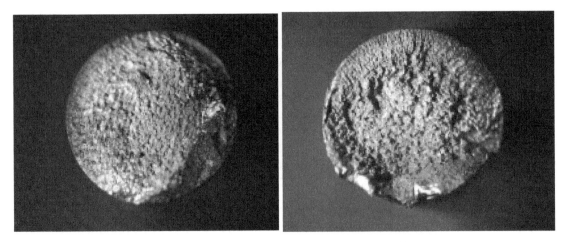

Fig. *7.14. Optical microscope photographs of specimens fatigued under S = 85.10 ksi.*

commenced before crack propagation occurred through the specimen cross-section, apparently because the fracture toughness of the material was reached at a crack length less than the specimen diameter.

Moreover, it should be noted that outer regions of the cross-section are subjected to a higher stress amplitude than inner regions, implying higher stress intensity factor range ΔK values (i.e., much faster crack growth) at the specimen edge. Therefore, crack initiation is more likely to occur at the specimen edge than internal defects. When the crack had propagated by a significant distance through the cross section of the specimen, $\Delta K \rightarrow K_c$ and fracture of the specimen occured instantaneously.

7.4. REFERENCES

[1] Dowling, N. E., *Mechanical Behavior of Materials*, 3rd ed., Prentice-Hall, Englewood Cliffs, NJ, 1998.

[2] Raske, D. T., and Morrow, J., "Mechanics of Materials in Low Cycle Fatigue Testing," ASTM STP 465, 1969, pp. 1–32.

[3] ETBX Strain-Life Fatigue Analysis Module (http://www.fea-optimization.com/ETBX/ strainlife_help.html, 2010).

[4] Manson, S. S., and Halford, G. R., "Practical Implementation of the Double Linear Damage Rule and Damage Curve Approach for Treating Cumulative Fatigue Damage," *International Journal of Fracture*, Vol. 17(2), 1981, pp. 169–192.

[5] Massing, G., "Residual Stress and Strain Hardening of Brass," *Proceedings of 2nd International Congress on Applied Mechanics*, Zurich, 1926.

[6] International Standard, ISO 1143-1975 (E): *Metals–Rotating Bar Bending Fatigue Testing*.

7.5. ASSIGNMENT

The objective of this lab is to perform stress-control fatigue tests according to the ISO standard code, examine the fatigue properties of steel specimens under different stress amplitudes, and use optical microscopy to study how crack growth was affected by the loading conditions. Fatigue tests will be performed with straight shank specimens consisting of AISI 1045 steel.

You are asked to carry out the following experiments, supplemented by analysis and discussion:

1. Perform fatigue tests with AISI 1045 steel straight shank specimens at three stress amplitudes corresponding to three different applied loads. (Note: Use the *S-N* data in Appendix A to select appropriate stress amplitudes in order to ensure the completion of all tests within the allowed time.)

2. Tabulate experimentally determined fatigue life versus stress amplitude results and fatigue life data calculated from Eq. (7.12). Compare experimental and calculated fatigue life data and discuss possible reasons for any observed differences. In particular, discuss the effect of stress amplitude on fatigue life.

3. Use an optical microscope to obtain photographs of cross-sections of fatigued specimens and discuss the underlying reasons for differences in cross-section texture and morphology observed at different stress amplitudes.

Applied load (lb)	Stress amplitude (ksi)	Calculated fatigue life	
		Cycles (N_f)	Time (min)
9.7	63.50	1,084,365	180.73
9.8	64.15	953,885	158.98
9.9	64.80	840,199	140.03
10.0	65.46	741,006	123.50
10.1	66.11	654,342	109.06
10.2	66.77	578,522	96.42
10.3	67.42	512,102	85.35
10.4	68.08	453,842	75.64
10.5	68.73	402,676	67.11
10.6	69.39	357,683	59.61
10.7	70.04	318,071	53.01
10.8	70.69	283,155	47.19
10.9	71.35	252,342	42.06
11.0	72.00	225,119	37.52
11.1	72.66	201,041	33.51
11.2	73.31	179,720	29.95
11.3	73.97	160,820	26.80
11.4	74.62	144,049	24.01
11.5	75.28	129,151	21.53
11.6	75.93	115,903	19.32
11.7	76.59	104,111	17.35
11.8	77.24	93,604	15.60
11.9	77.90	84,233	14.04
12.0	78.55	75,867	12.64
12.1	79.20	68,391	11.40
12.2	79.86	61,705	10.28
12.3	80.51	55,719	9.29
12.4	81.17	50,355	8.39
12.5	81.82	45,545	7.59
12.6	82.48	41,227	6.87
12.7	83.13	37,348	6.22
12.8	83.79	33,860	5.64
12.9	84.44	30,722	5.12
13.0	85.10	27,895	4.65

*Rotational speed $\approx$ 6,000 rpm; D = 0.146 in.; L = 4 in.; N_f estimated from Eq. (7.12)

Chapter 8: Wear Testing

8.1.1. Basic terminology

Lubricant:	A substance interposed between two surfaces to reduce friction and wear.
Wear:	Surface damage due to the removal of material during sliding contact.
Coefficient of friction:	Ratio of tangential (friction) force generated during relative surface movement to normal force transferred through the contact interface.
Seizure resistance:	The ability of the bearing material to resist physical joining (friction welding) when direct metal-to-metal contact occurs. High seizure resistance is important when the bearing operates in the mixed lubrication regime (e.g., start/stop operation phases, oil starvation, excessive clearance, and high roughness of bearing surface).

8.1.2. Lubrication Regimes

Sliding friction can be significantly reduced by introducing a lubricious substance between the rubbing surfaces. Friction in the presence of a lubricant is characterized by the formation of a thin film of pressurized lubricant (squeeze film) between the bearing surfaces. The ratio of the squeezed film thickness h to the average surface roughness R_a determines the lubrication regime of the bearing system.

8.1.2.1. Boundary Lubrication

Direct contact between high summits (asperities) of bearing surfaces occurs in the boundary lubrication regime. This is the most undesirable operation regime because of the extremely close proximity of the rough surfaces ($h < R_a$) that causes asperity-asperity contact interactions, resulting in high friction, material removal (wear), and possibly seizure. Severe engine bearing failures are mostly encountered under boundary lubrication conditions, favored under low relative speeds (i.e., start/stop cycles) and high loads. Special additives, known as extreme pressure (EP) additives, are incorporated in the lubricant to prevent seizure by forming a sacrificial boundary film conformal to the sliding metal surfaces.

8.1.2.2. Mixed Lubrication

Intermittent contact between the sliding surfaces occurs at moderate speeds and intermediate loads. Under these conditions, only a few asperity interactions are encountered, and a significant fraction of the normal load is supported by pressurized (hydrodynamic) fluid trapped within the valleys of the rough surfaces ($h \approx R_a$). Hence, mixed lubrication is referred to as the intermediate lubrication regime involving both boundary and hydrodynamic lubrication conditions.

8.1.2.3. Hydrodynamic Lubrication

High relative speeds and/or light loads favor complete separation of the moving surfaces by a continuous fluid film. Because the thickness of the hydrodynamic film is much larger than the surface roughness ($h >> R_a$), asperity interactions and, in turn, high friction and wear do not occur in this lubrication regime.

The above mentioned lubrication regimes can be observed in the so-called Stribeck curve, shown by Fig. 8.1, which is a plot of the friction coefficient μ versus the dimensionless parameter $\eta v/L$, where η is the dynamic viscosity of the fluid (gas) film between the moving surfaces, v is the relative sliding speed, and L is the normal load. As shown by the Stribeck curve, μ increases rapidly as $\eta v/L$ decreases in the mixed lubrication regime. This yields in a significant increase in wear and dissipation of mechanical work in the form of frictional heating. Thus, mixed lubrication is considered to represent transient lubrication conditions. The increase in μ with $\eta v/L$ in the hydrodynamic lubrication regime is due to the increase in thickness of the sheared viscous medium (liquid or gas).

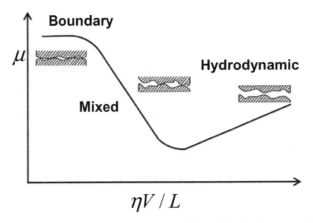

Fig. *8.1. Schematic of the Stribeck curve showing boundary, mixed, and hydrodynamic lubrication regimes.*

8.1.3. Determination of Friction Coefficient

The following parameters are either given or must be measured:
$D_1 =$ horizontal distance from the load cell to the center of rotation (7.62 cm)
$F_1 =$ horizontal force measured by the load cell
$D_b =$ diameter of ball bearing specimens used in the experiments (1.27cm)
$F_d =$ horizontal force applied to each contact point
$\mu =$ friction coefficient
$L_d =$ normal load applied at the contact point
$L =$ total normal load

By definition,

$$\mu = \frac{F_d}{L_d}$$

(8.1)

For the given geometry, it follows that

$$\mu = \frac{2\sqrt{2}F_1 D_1}{D_b L}$$

(8.2)

8.1.4. Seizure Point

Although the physical definition of seizure is well known, its identification during testing is more complex because surface "cold welding" cannot be observed while the test is running. Moreover, the four-ball tester, shown schematically in Fig. 8.2, has not been designed for extreme-pressure tests (ASTM D2266 [1]) that lead to gross seizure. As a result, seizure cannot be observed with such lubricants as mineral oil or engine oil under loads as high as 80 kg. More testing time is probably needed to observe seizure under such loads using these oils. In addition, it should be pointed out that an abrupt increase in friction force does not always imply seizure. It is not uncommon to observe a temporary increase in friction force by a factor of 2–2.5 during testing. Such an increase is attributed to a change in wear rate. Seizure was observed with the tested ball bearings that you are asked to observe under the microscope and compare with the ball bearings to be tested.

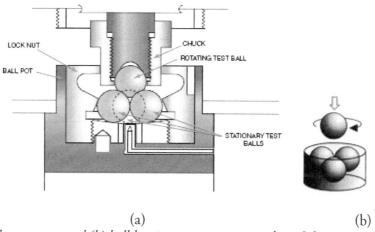

(a)   (b)

Fig. *8.2. (a) Four-ball wear tester and (b) ball bearing arrangement in oil cap [2].*

8.2. Equipment and Experimental Materials

- Four-ball wear tester (Falex).
- Optical microscope (Olympus SZ series).
- Ball bearing specimens (Falex) of 12.7 mm diameter consisting of AISI standard steel No. E52100, grade 25 EP of Rockwell C hardness equal to 65.
- Test conditions: 75°C test temperature and 1800 rpm rotational speed.

8.3. ASTM STANDARDS

8.3.1. Description

Three ASTM standards are available for the four-ball wear tester:
(a) ASTM D4172–94: *Standard Test Method for Wear Preventive Characteristics of Lubricating Fluid* [3].
(b) ASTM D2266–01: *Standard Test Method for Wear Preventive Characteristics of Lubricating Grease* [1].
(c) ASTM D5183–05: *Standard Test Method for Determination of the Coefficient of Friction of Lubricants* [4].

8.3.2. ASTM recommendations

Table 8.1 summarizes the parameters recommended by each ASTM standard. The ASTM protocols have been modified to fit the current machine and lab environment (Appendix 8A).

Table 8.1. ASTM standard parameters.

	ASTM D4172 (Wear)		ASTM D5183 (Friction)		
	Value	Error	Value	Error	Comment
Temperature (°C)	75	2	75	2	–
Speed (rpm)	1200	60	600	60	–
Duration (min)	60	1	60	1	10 min intervals
Load (N)	392	2	392	–	break
			98.1	–	minimum
			176	–	maximum

8.4. REPRESENTATIVE RESULTS

The two type of lubricants tested can be classified as poor lubricant (mineral oil) and performance lubricant (SAE engine oil-Valvoline premium). Two types of tests were performed to compare the wear protection provided by the mineral oil and the SAE 30 engine oil. One test, referred to as the *low-wear test*, was run under a light load (30 kg) for a relatively long time (25 min) at 1800 rpm, whereas another test, referred to as the *high-wear test*, was run under a relatively high load of 80 kg for 12 min. These tests were repeated five times to obtain an accurate statistical description of each oil behavior.

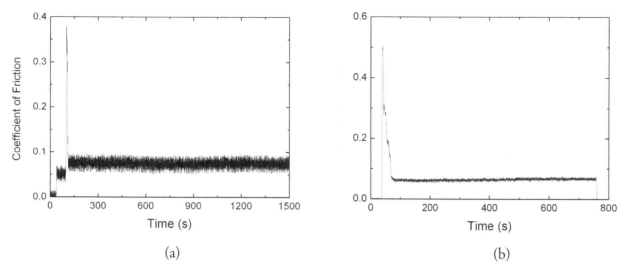

Fig. *8.3. Friction coefficient versus time for (a) low-wear and (b) high-wear tests performed with mineral oil. (Note the scale differences.)*

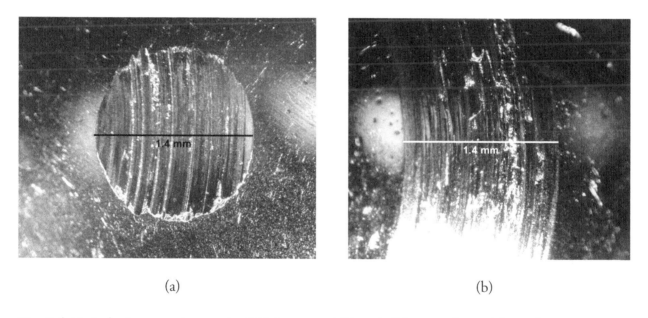

Fig. *8.4. Optical microscope photographs of (a) bottom and (b) top ball bearings obtained from a low-wear test performed with mineral oil.*

8.4.1. Mineral Oil

Figure 8.3 shows the friction coefficient versus test time from representative low-wear and high-wear tests performed with mineral oil. A peak can be observed during the transition phase, which is typical of the run-in process. The steady-state coefficient of friction is almost independent of applied load; however, the friction peak during the run-in phase is higher for the higher load. As expected, the wear rate was also found to increase with the load, as confirmed by photographs of the wear scars formed on the sliding steel surfaces.

Figure 8.4 shows wear scars on one of the three bottom bearings (Fig. 8.4(a)) and the top bearing (Fig. 8.4(b)) used in a low-wear test of Fig. 8.3(a). Because the rotating top bearing was in point-contact with each of the three stationary bottom bearings, the wear scar of the top bearing is circular.

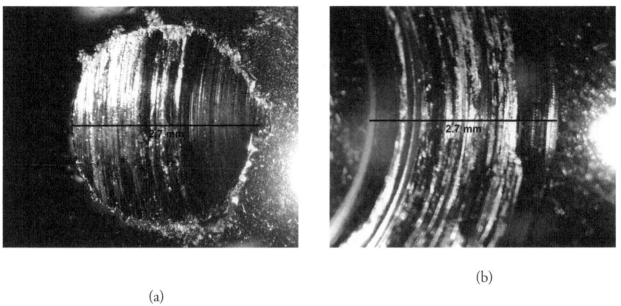

(b)

(a)

Fig. *8.5. Optical microscope photographs of (a) bottom and (b) top ball bearings obtained from a high-wear test performed with mineral oil.*

Figure 8.5 shows wear scars on one of the bottom bearings (Fig. 8.5(a)) and the top bearing (Fig. 8.5(b)) used in the high-wear test of Fig. 8.3(b). The average wear scar diameter in the low- and high-wear tests was found equal to 1.4 and 2.7 mm, respectively. Therefore, an increase in wear scar diameter was observed with an increase in load. The wear scar features were also affected by the load increase, revealing a transition from shallow to deep groove formation.

8.4.2. Engine Oil

Figure 8.6 shows friction coefficient versus test time plots from representative low-wear and high-wear tests performed with engine oil (SAE 30). A peak friction coefficient was observed in the run-in phase of the high-wear test (Fig. 8.6(b)), as opposed to the low-wear test (Fig. 8.6(a)) that did not reveal the occurrence

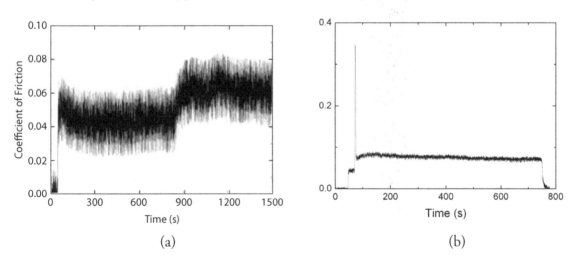

(a)   (b)

Fig. *8.6. Friction coefficient versus time for (a) low-wear and (b) high-wear tests performed with SAE 30 engine oil. (Note the scale differences.)*

of a peak friction coefficient. A comparison of Figs. 8.6(a) and 8.6(b) shows that the steady-state friction coefficient increased with the load.

Figures 8.7 and 8.8 show optical micrographs of wear scars formed on bearing surfaces tested under low- and high-wear conditions, respectively, in the presence of SAE 30 engine oil. The average wear scar diameter in the low- and high-wear tests was found equal to 0.4 and 2.0 mm, respectively. The wear scar features also changed as the load was increased, demonstrating a transition from shallow to deep grooves.

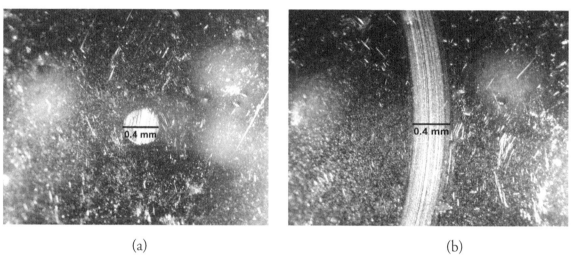

(a)   (b)

Fig. *8.7. Optical microscope photographs of (a) bottom and (b) top ball bearings obtained from a low-wear test performed with SAE 30 engine oil.*

(a)   (b)

Fig. *8.8. Optical microscope photographs of (a) bottom and (b) top ball bearings obtained from a high-wear test performed with SAE 30 engine oil.*

8.4.3. Comparison of friction and wear properties

Table 8.2 shows a comparison between friction coefficient and wear-scar diameter results obtained with mineral oil and SAE 30 engine oil under relatively low (30 kg) and high (80 kg) loads. The friction coefficient values given in Table 8.1 were calculated as the average of the steady-state friction coefficients obtained after the run-in phase. It can be seen that the wear scar diameters corresponding to SAE 30 engine oil are consistently less than those obtained with mineral oil under the same load, revealing the

Table 8.2. Friction and wear properties versus type of lubricant and normal load.

| Lubricant | Load (kg) | Scar diameter (mm) | | Friction coefficient |
		Top ball	Bottom ball	
Engine oil (SAE 30)	30	0.4	0.4	0.043
Mineral oil	30	1.4	1.4	0.078
Engine oil (SAE 30)	80	2.0	2.0	0.076
Mineral oil	80	2.7	2.7	0.070

superior wear resistance (protection) provided by SAE 30 engine oil relative to mineral oil. For the SAE 30 engine oil, increasing the load from 30 to 80 kg resulted in the transition from mixed to boundary lubrication conditions, as evidenced by the significant increase in friction coefficient (by ~43%). However, in the case of the mineral oil, sliding occurred in the boundary lubrication regime under both loads, as indicated by the similar friction coefficients obtained under 30 and 80 kg loads. This is another illustration of the superior lubrication and protection characteristics of the SAE 30 engine oil.

8.5. REFERENCES

[1] Annual Book of ASTM Standards, D2266–01: *Standard Test Method for Wear Preventive Characteristics of Lubricating Grease (Four-Ball Method).*
[2] South Research Institute, *Four-ball Friction and Wear Tests.*
[3] Annual Book of ASTM Standards, D4172–94 (2004): *Standard Test Method for Wear Preventive Characteristics of Lubricating Fluid (Four-Ball Method).*
[4] Annual Book of ASTM Standards, D5183–05: *Standard Test Method for Determination of the Coefficient of Friction of Lubricants Using the Four Ball Wear Test Machine.*

8.6. ASSIGNMENT

The main objectives of this lab are to learn the principles of lubricant testing under the ASTM code and obtain some insight into sliding wear mechanisms. In particular, you will learn about different lubrication regimes, seizure mechanisms, lubricant behavior, and effect of test load, temperature, and sliding speed on the friction and wear characteristics of steel ball bearings. You will also learn how to design an experiment from a statistical perspective and how to perform wear experiments using a state-of-the-art four-ball tester.

Your report should address and provide explanations to the following:

1. Examine the previously tested bearings with an optical microscope. Discuss the wear features observed on the wear tracks and characterization of the wear mechanism. These bearings were tested under a relatively low load of 30 kg for 25 min using two different lubricants: SAE 30 (engine oil) and mineral oil.

2. Measure the wear-scar diameter and discuss the lubricant effect on wear damage of each ball.
3. Perform four-ball experiments with the same lubricants (i.e., mineral oil and SAE 30) under a high load (80 kg) for a relatively short duration (12 min).
4. Repeat (1) and (2) using the ball bearings that you tested in (3).
5. Discuss and explain differences between all bearing specimens that you examined. In particular, evaluate the effects of load, time, and lubricant on the friction and wear behavior of the steel bearings.

NOTE: THE ASTM PROTOCOLS HAVE BEEN MODIFIED TO FIT THE PARTICULAR WEAR TESTING MACHINE AND LABORATORY ENVIRONMENT.

Four-ball tester

Cleaning Procedure

- Make sure you wear new gloves.
- Thoroughly clean the four balls, collet, and entire ball cup assembly by first soaking in isopropyl alcohol for 30 s.
- Drain and rinse with fresh isopropyl.
- After cleaning, it is important to handle all parts using a fresh wipe. No trace of cleaning fluid should remain when the test oil is introduced and the machine assembled.

Set-up Procedure

- After the removal of the cup and the pivot pin, the support arm (1) should hold the lever arm down and away from the air bearing.
- Move belt to appropriate pulley for desired rpm for test. For the top pulley set it at 1800 rpm, for the middle pulley set it at 1200 rpm, and for the lower pulley set it at 600 rpm.
- Insert one of the clean test balls (top ball) into the ball collet (do not handle ball with bare hands). Insert the ball collet into the spindle of the test machine (after cleaning the spindle) and tighten the screw on top of the machine, hand tight.
- Assemble three of the clean balls in the test-oil cup and lock the balls in position by hand tightening the lock-nut into the ball cup. Then, using the torque wrench, tighten the collar to 25 ft/lbs (see scale on wrench).
- Install test-oil cup into the heater assembly and ensure that it locks into the detent in the bottom of the assembly.
- Pour the lubricant to be tested into the test-oil cup to a level of at least 3 mm (1/8 in) above the top of the balls. Ensure that this oil level still exists after the lubricant fills all the voids in the test oil cup assembly.
- Install the ball assembly on top of the air bearing and move support arm (2) under air bearing.
- Install pivot pin under air bearing at "x" marks.
- Swing support arm (1) back against machine away from the lever arm.
- Slightly lower the left side of the lever arm and swing support arm (2) back against the machine and away from the air bearing.
- Attach the thermocouple and heater plugs to the cup assembly.
- If applicable, close safety cover.
- Apply the desired load to the lever arm by adding weights to the hanging rod. Apply the test load slowly to avoid impact loading.

- Rotate the air pressure valve to the "on" position, and make sure the gauge reads 12–15 psi. This can be adjusted using the regulator knob.
- Turn key to the "on" position to turn on the main power (see tag).
- The temperature controller is preset to maintain a temperature of 75°C.
- Once the air bearing is on, set the load cell at 0 (by pressing the "TARE" button) after making sure that the cup holder is not in contact (make sure all parts and test oil are in place).
- Press the "load cell reset" button to the right of the load cell controller. The "force ok" green led will illuminate.
- Set desired duration of test on timer controller in seconds.
- Turn on the heater switch. Heater will not turn on until the "start" button is pressed.
- Set the motor switch to *'auto'* (see tag).
- When you are ready, press the "start" button. Once the button is pressed, the "test running" green led will illuminate.
- The heater will start to heat the sample and the "heater on" light will illuminate.
- Once the set temperature is achieved, the "up to temp" green led will illuminate.
- At this point, the timer will start to count down and the motor will start running. As soon as the time set in the timer is attained, the "timer running" green LED will illuminate.
- Once the timer is completed, the motor and heater will automatically stop. The "test complete" red LED will illuminate. To stop the test before the time set in the timer is reached, press the "emergency stop" button.
- Turn the motor switch to the "off" position and turn off the heater switch.
- Once the sample has cooled down such that it could be touched, it can be removed and examined in accord to the ASTM standard.
- Once the thermocouple has cooled down, the "not up to temp" LED will illuminate and the timer will be reset and ready for the next test.
- Perform a new test following the above procedure.

Microscope

- Important: Leave the three balls clamped. Drain the test oil from the three-ball assembly and wipe out the scar area with a tissue.
- Set the assembly on a special base of the microscope that has been designed for this purpose.
- Make two measurements on each of the wear scars. Take one measurement of the scar along a radial line from the center of the holder. Take the second measurement along a line 90° from the first measurement.
- Report the arithmetic average of the six measurements as the scar diameter expressed in units of millimeters.
- If the average of the two measurements on one ball varies from the average of all six readings by more than 0.1mm, investigate the alignment of the top and bottom balls.
- Repeatability: In about one case out of twenty, the repeatability may exceed 0.4 mm in scar diameter difference.

Appendix A: Report Template[1]

Mechanical Behavior of Materials (ME108)

UC BERKELEY — MECHANICAL ENGINEERING

[Lab Title Here]

[Student names]

Date Submitted:
Date Performed:
Lab Section:
Lab GSI:

ABSTRACT (MAXIMUM 150 words)

The abstract is a *clear, concise, and complete* summary of the project, including the purpose, experimental procedure, results, and major conclusions. The abstract should be one or possibly two paragraphs in length (≤ 150 words) and must appear on the title page. The abstract should provide the scope, approach, and main findings but not any detailed or specific information given in the main body of the report. The abstract *is not* an introductory section of the report, and must be able to stand alone.

Question to address:	How to address it:
What is the report about, in miniature and without specific details?	**State main objectives and theory**. (What did you investigate? Why?) **Describe methods**. (What did you do?) **Summarize the most important results**. (What did you find out?) **State major conclusions and significance**. (What do your results mean?)

Additional tips:
Write this section last, extract key points from each section, condense in successive revisions.

What to avoid:
Do not include references to figures, tables, or sources, and trivial information.

---

[1]This lab template was modified from the templates/guidelines developed by Northeastern University's Department of Mechanical and Industrial Engineering (http://www1.coe.neu.edu/~adams_g/MIM-U455/Template.doc.), the writing center at the University of Wisconsin-Madison (http://writing.wisc.edu/Handbook/SciRep_Disc.html), Eli Patten's UC Berkeley ME Department 107B handouts, and Professor Lisa Pruitt's reference guide.

The introduction may have some elements that resemble the abstract, but must include other information as well. Do not worry about anything that seems repetitive with the abstract—they are distinct entities within the report. The introduction serves, literally, to introduce the reader to the report that follows. It needs to draw the reader in, to give the reader context, to make the reader interested, and to ensure that the relevance of the report is apparent.

Do not write verbatim what was written in the particular chapter of a book (you may write in your own words and cite the book)! The document should be 1.5 spaced in 11 point Times New Roman font. The margins should be 1 in. on all sides. You can choose to justify the text or not, but whatever you decide, be consistent. The Introduction section should include the pertinent literature and theory from referenced sources (text books should be referenced). These references must be listed in the Reference section (See Appendix B for more information about the recommended format of the references).

Questions to address:	How to address them:
What is the purpose?	**Precisely state** the experimental objective(s). **State** the scope of the report, including any limitations of what it does or does not offer to the reader. This information should be included in the closing paragraph of the introduction.
Why is it important?	**Motivation** for this study and/or application of the results (*relevance*).
What is the theory you are using?	Briefly **describe** the *theory*: The background necessary for understanding the experimental method, reduction of observations, and the presentation and discussion of the results. Quickly explain all the fundamentals behind the equations and principles used. **Summarize** relevant theory to provide context, key terms, and concepts so the reader can understand the experiment.

Additional tips:

1. Process: this is often the hardest section to write—spend time on the introduction, use words and sentences that orient readers to the report topic, purpose, and theory.
2. Move from general to specific: problem in real world → your experiment.
3. Engage your reader: answer the questions, "What did you do?" and "Why should I care?"
4. Be selective, not exhaustive, in choosing the amount of detail to be included.

This section should include the background theory necessary to understand the experimental method, reduction of observations, and presentation and discussion of the results. Quickly explain all the fundamentals behind the equations and principles you will use. The very basic background information found in commonly used texts should be referenced (see Appendix B).

Question to address:	How to address them:
What is the theory you are using?	Briefly **describe** the *theory*: the fundamentals behind the equations and principles you will use. **Summarize** relevant theory to provide context, key terms, and concepts so that your reader can understand the experiment.

Additional tips:
1. Ensure that the links between your experiment and theory are clearly stated.
2. Type all equations.
3. Define all variables, with units, in the equations you use.

Description of the laboratory setup: Because you will not have space for every single detail, you need to be selective and use your judgment on what is most important and, therefore, worthy to include.

Describe the main components of the experimental instrumentation: company name, model, accuracy, range, and any other applicable specifications.

Description of the experimental procedure: Discuss what was manipulated (independent variables), what was observed (dependant variables), what was held constant (controls), and other measurements obtained (such as ambient temperature). Include where and how (including sampling frequency and duration) these data were measured (collected) and the instrument(s) used. This section should also include a concise description of operation and important measurement techniques.

Questions to address:	How to address them:
How did you study the problem?	Briefly **explain** the general procedure you used, including what was manipulated (**independent variables**) and what was observed (**dependant variables**).
What did you use?	**Describe** what materials and equipment you used.
How did you proceed?	**Explain** the steps you took in your experiment.

Additional tips:
1. Provide enough detail for replication.
2. Order procedures chronologically.
3. Use past tense to describe *what you did*.
4. Quantify when possible (measurements, amounts, time, temperature, etc).

What to avoid:
1. Do not include trivial details.
2. Do not mix results with procedures.

This section is a formal, concise description of the tables, graphs, and final results that are needed to understand the discussion and conclusions. Be selective in what you present and try to distill it to what is most important. Just because you have the data or already made a graph does not mean that you need to present it (*if it doesn't help, it hurts*). Clarify what each table and graph is, point out its important features and characteristics, and compare the various results (where appropriate). Remember to refer your reader to specific Figures, Tables, and Appendices, where applicable, and show your calculations and data manipulation. Note that it is preferred to have Figures and Tables close to the text where they are first introduced. The goal here is to report the results—***not*** to discuss whether they are good or bad results. Usually the trends in a graph are pointed out, but not fully explained. The explanation of the trend or comparison should be saved for the Discussion section.

Question to address:	How to address it:

What did you observe?	For **each** experiment or procedure:
	Report the main result(s) in appropriate format:
	    tables (title, labels, units, caption)
	    graphs (title, labels, units, legend, caption)
	    text (be selective, concise, quantitative, DO NOT explain or discuss)

Additional tips:
1. Order the results logically: i.e., from most to least important.
2. Use past tense to describe *what happened*.
3. Be quantitative: "Yield strength of material A was lower than that of material B by 20%," instead of "Yield strength of material A was lower than material B."

What to avoid:
1. Do not simply repeat data in tables and graphs in the text; select and summarize, e.g., "Yield strength of material A was lower than that of material B by 20%," instead of "Yield strength of material A was 160 MPa, and yield strength of material B was 200 MPa."
2. Do not explain the results. That will be done in the Discussion section. (*Why* was the yield strength lower for material A? Did you expect the yield strength of material A to be less than that of material B? Why?)
3. Avoid extra words: "It is shown in Table 1 that X induced Y"→"X induced Y (Table 1)."

5. DISCUSSION (1 page)

In the discussion, you should point out how your experimental results compare with theory, and suggest and explain reasons for deviations. Do not ignore deviations in your data or unexpected results. Discuss the sources of error in this section.

Questions to address:	How to address them:
What do your observations mean?	**Summarize** the most important findings at the beginning.
What conclusions can you draw?	For **each** major result: **Describe** the patterns, principles, relationships your results show. **Explain** how your results relate to expectations and to theory cited in the Introduction. Do they agree, contradict? **Explain** reasons for contradictions, unexpected results (sources of error or discrepancy). **Describe** what additional testing might resolve the underlying reasons for these contradictions.
How do your results fit into a broader context?	**Suggest** practical applications of your results. **Extend** your findings to other situations. **Give** the big picture: do your findings help you understand the broader theory?

6. CONCLUSIONS (1 paragraph)

Summarize the major findings and how they relate to the theory in a paragraph. There may be repetition with the Abstract, but that is okay as long as different wording is used.

7. REFERENCES

At least 3 references, excluding the web, using the Vancouver format (see Appendix B) must be included in the report.

8. APPENDIX A: TYPES OF INFORMATION INCLUDED IN APPENDICES

Note: Each appendix should begin on a separate page. Appendices are used for specific calculations or data that would clutter the main body of the report. Remember to specifically refer the reader to an appendix in the body of the report. Label appendices with capital letters (A-Z) and include a descriptive title for each appendix.

1. Detailed derivations.
2. Raw data.
3. Computer code.
4. Other information that is too detailed to place in the main body of the report.

9. APPENDIX B: THE NUMBERING (VANCOUVER) SYSTEM FOR REFERENCES

- Numbers should appear in text as [1].
- Date comes at the end of the reference.
- [number] Author name, *Book Title*. (Place of publication: publisher, date).
- [1] D. A. Ratcliffe, *The Peregrine Falcon*, (London: Powser, 1993).

- Numbers should appear in the text as non-superscript numbers in square brackets, preceding punctuation. Where possible without losing clarity, the number should be placed at the end of a sentence or before an obvious break in punctuation.
- References should be numbered in the order in which they are cited in the text. If the same reference is used twice, do not give it two numbers. Use the first number assigned to that reference for the entire report.
- Names should not be inverted. The publication date should appear last for books, and preceding the page numbers for journal articles and chapters in edited volumes.

Books

1. D. A. Ratcliffe, *The Peregrine Falcon*, 2nd edn (London: Poyser, 1993).
2. R. Jurmain, H. Nelson & W. A. Turnbaugh, *Understanding Physical Anthropology and Archeology*, 4th edn (St Paul, MN: West Publishing Co., 1990).
3. J. A. Hazel, *The Growth of the Cotton Trade in Lancashire*, 3rd edn. 4 vols. (London: Textile Press, 1987–8).

Journal articles

1. J. H. Werren, U. Nur and C.-I. Wu, Selfish genetic elements. *Trends in Ecology and Evolution*, **3** (1988), 297–302.
2. S. W. Trimble, Streambank fish-shelter structures help stabilize tributary streams in Wisconsin. *Environmental Geology*, **32**:3 (1997), 230–4.

Chapters in edited books

1. N. M. Simmons, Behaviour. In *The Desert Bighorn*, ed. G. Monson and L. Summer. (Tucson, AZ: University of Arizona Press, 1980), pp. 124–44.

Appendix B:
Rubric For Assessing Lab Reports

(Adapted from North Carolina State University and Union College lab assessments)

Section	Assessment				Score (1–4)
	Beginning or Incomplete (1 point)	**Developing (2 points)**	**Accomplished (3 points)**	**Exemplary (4 points)**	
Abstract	Several major aspects of the experiment are missing, lack of understanding about how to write an abstract, too long	Misses one or more major aspects of carrying out the experiment or the results, too long	References most of the major aspects of the experiment, some minor details are missing	Contains reference to all major aspects of the experiment and the results, well-written	
Introduction and Theory	Very little background information/theory provided or information is incorrect	Some introductory information/theory, but still missing some major points	Introduction is nearly complete, missing some minor points	Introduction complete and well-written; provides all necessary background/ theory	
Experimental Procedure	Missing several important experimental details, materials, variables, or not written in paragraph format	Written in paragraph format, still missing some important experimental details, materials, variables	Written in paragraph format, important experimental details are covered, some minor details missing	Well-written in paragraph format, all experimental details are covered	
Results	Graphs, tables contain errors or are poorly constructed, missing titles/captions/ units, text includes discussion or repeats the data.	Most graphs, tables OK, some still missing some important or required features, text includes discussion or repeats the data.	All graphs, tables are correctly drawn, but some have minor problems or could still be improved. Text is quantitative.	All graphs, tables are correctly drawn, contain titles/captions/ labels, referenced in the text. Text is quantitative and selective.	

Section	Assessment				Score (1–4)
	Beginning or Incomplete (1 point)	Developing (2 points)	Accomplished (3 points)	Exemplary (4 points)	
Discussion	Very incomplete or incorrect interpretation of trends and comparison of data indicating a lack of understanding of results	Results correctly interpreted and discussed; partial but incomplete understanding of results is still evident	Almost all of the results have been correctly interpreted and discussed, there is a link between data and theory, only minor improvements are needed	All important trends and data comparisons have been interpreted correctly and discussed; experiment is discussed in context of theory, good understanding of results is conveyed	
Conclusions	Conclusions missing or missing the important points	Conclusions regarding major points are drawn, but many misstated, lack of understanding	All important conclusions have been drawn, could be better stated	All important conclusions have been clearly made, student shows good understanding	
References and Appendices	Fewer than 3 references (excluding web); incorrectly formatted references or appendix	At least 3 references, incorrectly formatted references or appendix	At least 3 correctly formatted references, correct appendices, not referenced correctly in text	At least 3 correctly formatted references, correct appendices, all referenced correctly in text	
Spelling, Grammar, and Sentence structure	Frequent grammar and/or spelling errors, writing style is rough and immature	Occasional grammar/spelling errors, generally readable with some rough spots in writing style	Less than 5 grammar/spelling errors, mature, readable style	All grammar/spelling correct and very well-written	
Appearance and Formatting	Sections out of order, too much handwritten copy, sloppy formatting, obviously written by multiple authors	Sections in order, contains the minimum handwritten copy, formatting is rough but readable	All sections in order, formatting generally good but could still be improved	All sections in order, well-formatted, very readable. Cohesively written in one voice. Professional quality.	
				Title Page: Correct +2	
				Length: At or below limits +2	
				Total (40 maximum)	

Title of Lab:_____

One spreadsheet for each lab report and one summary spreadsheet containing the totals for the entire semester

Item	Students					
	1	2	3	......etc.	Average	% of students with 3 or 4
Abstract/Summary						
Introduction & Theory						
Experimental Procedure						
Results: data, figures, graphs, tables, etc.						
Discussion						
Conclusions						
Spelling, grammar, sentence structure						
Appearance and formatting						
TOTAL for each student						

CPSIA information can be obtained
at www.ICGtesting.com
Printed in the USA
FSOW03n1840270815
10456FS